Eagan Press Handbook Seri

Fats and Oils

Clyde E. Stauffer

eagan·press
St. Paul, Minnesota, USA

Library of Congress Catalog Card Number: 96-83465
International Standard Book Number: 0-913250-90-2

Printed in the United States of America on acid-free paper

American Association of Cereal Chemists
3340 Pilot Knob Road
St. Paul, Minnesota 55121-2097, USA

About the Eagan Press Handbook Series

The Eagan Press Handbook series was developed for food industry practitioners. It offers a practical approach to understanding the basics of food ingredients, applications, and processes—whether the reader is a research chemist wanting practical information compiled in a single source or a purchasing agent trying to understand product specifications. The handbook series is designed to reach a broad readership; the books are not limited to a single product category but rather serve professionals in all segments of the food processing industry and their allied suppliers.

In developing this series, Eagan Press recognized the need to fill the gap between the highly fragmented, theoretical, and often not readily available information in the scientific literature and the product-specific information available from suppliers. It enlisted experts in specific areas to contribute their expertise to the development and fruition of this series.

The content of the books has been prepared in a rigorous manner, including substantial peer review and editing, and is presented in a user friendly format with definitions of terms, examples, illustrations, and trouble-shooting tips. The result is a set of practical guides containing information useful to those involved in product development, production, testing, ingredient purchasing, engineering, and marketing aspects of the food industry.

Acknowledgment of Sponsors for *Fats and Oils*

Eagan Press would like to thank the following companies for their financial support of this handbook:

AC HUMKO Corporation
Memphis, Tennessee
800/691-1106

Ag Processing (AGP)
Omaha, Nebraska
402/498-2281

Archer Daniels Midland
Decatur, Illinois
800/637-5866

Bunge Foods
Bradley, Illinois
800/828-0800

C&T Quincy
Richmond, Virginia
800/284-6457

Quest International—Specialty Lipids Division
Hoffman Estates, Illinois
800/621-4710

Eagan Press has designed this handbook series as practical guides serving the interests of the food industry as a whole rather than the individual interests of any single company. Nonetheless, corporate sponsorship has allowed these books to be more affordable for a wide audience.

Contents

Fats and Oils

Functional Properties of Fats and Oils

Functions in Food Systems

Fats and oils have always been an integral part of the human diet. Of prime importance is their role as a calorie-dense food component—they have nine kilocalories per gram versus four kilocalories per gram for starch or protein. The present-day concern with obesity and high fat content is actually a historical and geographical anomaly. For the greater part of human existence, the search for adequate food energy sources occupied a large segment of time, and fat was a prized component of the diet.

The varied cultural preferences for traditional food fats spring, in part, from geographic roots. In colder climates (e.g., northern Europe) most fats were derived from animal sources, and lard, tallow, and butter were the main fats. Thus, *plastic shortenings* were used in most culinary recipes that were brought to North America by northern European immigrants. In warmer climates (e.g., Mediterranean countries), vegetable oils, such as olive and sesame seed oils, were the predominant available fats; the cuisine of these countries reflects this difference. More recent times have seen a cross-cultural interchange in this respect, but to some extent the traditional patterns still exist.

It is difficult to adequately condense many large tables of data on the availability and consumption of fats and oils, but one example can be instructive. The annual per capita (adult) consumption of fats and oils in the United States, as of the late 1980s, was approximately:

- 24 lb of salad and cooking oils
- 23 lb of bakery shortening and frying fats
- 20 lb of fat from meat, poultry, fish, and cheese
- 13 lb from butter, margarine, and other miscellaneous sources.

This total of 80 lb per person per year represents about 38% of the calories in the diet. With the recent push to reduce this number to 30%, the above numbers will probably decrease, but the proportions will likely remain about the same.

Commercial (as opposed to home) use of fats and oils accounts for most of the shortening and frying fat category, some part of the *salad oil* category, fats for margarine production, and many of the miscellaneous items (confectionery coatings, oil-based whipped toppings, coffee

In This Chapter:

Functions in Food Systems
  Functions in Processing
  Sensory Functions
  Nutrition

Functional Properties
  Chemical Structure and Fat Properties
  Functional Characteristics
  Chemical Reactions
  Fat and Oil Sources and Compositions

Plastic shortening—A firm fat that contains solid fat crystals surrounded by oil. The consistency or ability to be shaped and molded is related to type of fat, type of crystals, and temperature.

Salad oil—A refined liquid oil that does not cloud when stored under refrigeration conditions.

whiteners, etc.). The specifications for these fats and oils are generally more stringent than for fats and oils used in the home. Commercial equipment and processes are more sensitive to variations in fat characteristics; the specified properties for the finished product are more narrowly defined; and compensating for ingredient variation takes longer than in home cooking. The people responsible for efficient operation of such a plant must understand how fat and oil properties affect production and product characteristics and how to write and enforce specifications for these raw ingredients, so that the outcome is satisfactory both to the consumer and to the owners of the plant.

FUNCTIONS IN PROCESSING

Shortening has an important functional role in processing baked foods, which is discussed in more detail in Chapter 5. It provides structure in some products (for example, cookie dough) and lubrication in others. In cookie dough and cake batters, shortening holds the finely divided air bubbles that serve as the nuclei for leavening gases and give a fine grain to the finished product. In roll-in doughs (Danish, puff pastry), plastic fat prevents adjacent dough layers from knitting together during proofing, so the final product has a flaky, many-layered structure. Fat is the main structural element in the fillings and cream icings used in and on bakery sweet goods of various sorts.

Fat for frying is important in many segments of the food industry: potato chips and corn chips are deep-fried; nuts are roasted; cake doughnuts, potatoes, and battered or breaded poultry or fish are fried. These and other food items depend not only on the heat transfer properties of fat, but also on the flavor imparted by the fat. The proper choice and correct use of fat are necessary to obtain acceptable results in these products.

SENSORY FUNCTIONS

Flavor. Fat contributes flavor to foods. This is most apparent in fried foods, because absorbed fat becomes an integral part of the finished product. The flavor is the result of products of numerous reactions between fat and other food components such as proteins and carbohydrates. *Oxidation* products of fat are also involved, giving a situation that is extremely complex in terms of the number of possible flavor compounds present. Nevertheless, without these compounds, the product would not have the characteristics we have come to associate with fried foods. For example, when a taste panel assesses food fried in completely fresh (refined) oil, the typical comment is usually something like "tasteless."

Fat or oil can contribute flavor directly, either in a positive or negative sense. Olive oil, which has a unique flavor, is prized for gourmet salad dressings specifically because of the flavor notes it contributes. On the other hand, if a fat or oil is exposed to air and allowed to oxidize slightly, a flavor often develops that is referred to as a *"reversion flavor."* The characteristics of this flavor differ in different oils ("beany,"

Oxidation—Chemical reaction in which the double bond on a lipid molecule reacts with oxygen to produce a variety of chemical products. The consequences of this reaction are loss of nutritional value and formation of the off-flavors associated with rancidity.

Reversion flavor—Mild off-flavor developed by a refined oil when exposed to oxygen. Reversion occurs rather easily, and the off-flavor, while undesirable, is not as objectionable as rancidity caused by oxidation.

"grassy," and "metallic" are some descriptive terms used), but it is invariably considered to be negative in terms of product quality.

Texture. Several aspects of food texture (or mouthfeel) are attributable to fat. It tenderizes the food, making it easier to bite and chew. It also makes a food feel moister in the mouth. As an example of these two aspects, consider the texture of steak—compared to a well-marbled steak, the lean cut of meat is characterized as "tough" and "dry." Oil also lubricates the food particles, helping to clear them from mouth surfaces (teeth and palate) more readily.

These tenderizing and lubricating characteristics are primarily attributable to the liquid fraction of the fat. If the fat melting point is much higher than body temperature, the fat does not melt in the mouth and the residual solid portion gives a "waxy" mouthfeel, an undesirable situation. This is particularly noticeable when the fat is a major component, as in fillings, icings, and confectionery coatings.

NUTRITION

As mentioned above, a major contribution of fat to the diet is as a dense, easily stored source of calories. An example is pemmican, traditionally used by Native Americans as rations when on an expedition (either hunting or warfare). This was made by mixing shredded jerked meat and dried berries in a container, then pouring melted animal fat over the blend. While perhaps less appetizing than the products of the U.S. Quartermaster Corps, pemmican served the same purpose: a storage-stable, readily transportable source of food energy for consumption during periods of high energy expenditure.

Fat also makes other positive nutritional contributions. It carries fat-soluble vitamins (A, D, and E). Its component fatty acids are metabolized by the body into phospholipids, which are essential parts of cell membranes. Without the proper balance of saturated and unsaturated fats, the membranes are either too solid or too fluid, and cell integrity is lost. Finally, certain polyunsaturated fatty acids are the precursors of lipid hormones (*prostaglandins*) that are needed by the body. If the diet completely lacks these essential fatty acids, certain untoward symptoms (hair loss, scaly skin, loss of reproductive capability) appear.

Prostaglandins—A group of specialized lipids that play important metabolic roles in humans. They are formed in the body from dietary essential fatty acids.

Functional Properties

Before beginning the discussion of functional properties, certain commonly used terms should be defined:

- Fat is a natural lipid material that is more or less solid at room temperature.
- Oil is a similar material that is liquid at room temperature.
- *Shortening* (mainly a baking term) is a fat or oil that contains no water.
- *Margarine* is a fat containing up to 20% water as a water-in-oil emulsion.

Shortening—A type of fat used in baking or frying. The name comes from the ability to tenderize or "shorten" baked products.

Margarine—A product category similar to dairy butter in composition and color. It contains 80% fat, 16% water, and 4% other ingredients such as salt.

CHEMICAL STRUCTURE AND FAT PROPERTIES

Fatty acids—A group of chemical compounds characterized by a chain made up of carbon and hydrogen atoms and having a carboxylic acid (COOH) group on one end of the molecule. They differ from each other in the number of carbon atoms and the number and location of double bonds in the chain. When they exist unattached to other compounds, they are called free fatty acids.

Ester—The chemical linkage that holds an alcohol group (OH) and an acid group (such as COOH) together. An ester bond is the connection between a fatty acid and glycerol in glycerides.

Glycerol—A three-carbon chain, with each carbon containing an alcohol group. One, two, or three fatty acids may be attached to glycerol to give a mono-, di-, or triglyceride.

Triglyceride—Three fatty acids attached to a glycerol molecule. If the three fatty acids are the same, it is a simple triglyceride; if they are different from each other, it is a mixed triglyceride. Mixed triglycerides are the most common chemical components in fats and oils.

Fatty acids. Fats are *esters* of fatty acids and *glycerol*. Most fats occur in the form of *triglycerides*, in which three fatty acids are attached to the glycerol. Fatty acids contain the *carboxyl group* (COOH) and an *aliphatic* carbon chain of variable length (Boxes 1-1 and 1-2). The general formula is R-COOH, where R is the aliphatic group. With few exceptions, fatty acids are linear, range in size from four to 24 carbons, and contain an even number of carbons.

The chains can be *saturated* (having no double bonds), *monounsaturated* (one double bond), or *polyunsaturated* (two or more double bonds). Box 1-3 shows three fatty acids that each contain 18 carbons but have different numbers of double bonds and thus differ in saturation. In accordance with the Geneva system of nomenclature, the carbon atoms of fatty acid chains are numbered consecutively, starting with the carbon atom of the carboxyl group as number 1 and continuing to the carbon atom in the terminal methoxy group. A shorthand designation of fatty acids is often used, based on the number of carbon atoms in the molecule and the degree of unsaturation (i.e., the number of double bonds in the molecule). The most common fatty acids in edible fats and oils are those containing 16 or 18 carbon atoms. These include the saturated palmitic (C16:0) and stearic (C18:0) acids, the monounsaturated oleic acid (C18:1), and the polyunsaturated acids—linoleic acid with two double bonds (C18:2) and linolenic acid with three double bonds (C18:3). A list of natural fatty acids along with their common names, designations, and main sources is given in Appendix A at the end of the book.

Another naming convention is applied to unsaturated fatty acids that have a physiological function as prostaglandin precursors (see Chapter 9). These are the *omega* (ω) *fatty acids.* The number of carbon atoms between the double bond and the terminal methyl group is designated as ω plus a number. For example, the end of the linoleic acid chain is $CH_3(CH_2)_4 CH=CH-$, so linoleic acid is termed an ω6 fatty acid (Box 1-4). Similarly, oleic acid is C18:1ω9 and linolenic acid is C18:3ω3. The ω3 fatty acids have some unique nutritional properties, which are con-

---

Box 1-1. Structure of a Fatty Acid

This is a saturated fatty acid; it has no double bonds.

Aliphatic carbon chain

Terminal methyl group   Carboxyl group

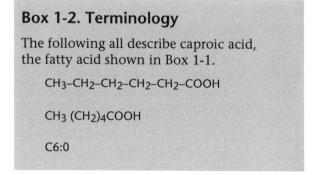

Fatty acid

Box 1-2. Terminology

The following all describe caproic acid, the fatty acid shown in Box 1-1.

$$CH_3-CH_2-CH_2-CH_2-CH_2-COOH$$

$$CH_3 (CH_2)_4COOH$$

C6:0

nected with their conversion to a particular group of prostaglandins (see Chapter 9).

The chemical reactivity of unsaturated fatty acids is determined by the position as well as the number of the double bonds in the molecule. Reactivity increases markedly with an increase in the number of double bonds, provided they are *conjugated* (separated only by one single bond) or *methylene-interrupted* (separated by a -CH₂- unit) (Box 1-5). If a fatty acid has two isolated double bonds (separated by two or more methylene units), its reactivity is only slightly greater than that of a fatty acid that has only one double bond. These differences are important when the fat is subjected to oxidation and also during the hydrogenation process.

In most naturally occurring unsaturated fatty acids, the double bonds are in the *cis* configuration. This means that the carbon chains on the

Carboxyl group—The chemical functional group on one end of a fatty acid. This is the same as a carboxylic acid (COOH), which can lose a proton and become COO⁻, or combine with an alcohol group to form an ester.

Aliphatic—Describing a straight chain of carbons with no branching or ring structure.

Saturated—Describing a carbon chain in which the carbons are connected to each other by single bonds, drawn as C–C. It has no carbon-to-carbon double bonds.

Monounsaturated—Describing a fatty acid that has one double bond (C=C) in the carbon chain. Oleic acid is the most common of these.

Polyunsaturated—Describing a fatty acid that has more than one double bond (C=C) in the carbon chain. Linoleic acid is an example.

Omega fatty acids—A method of nomenclature that designates the number of carbons between the terminal -CH₃ group and the last double bond in the fatty acid. This is useful in discussing the physiological role of certain polyunsaturated fatty acids.

Conjugated—Describing a situation in which double bonds between carbon atoms occur in a series with one single bond in between (C=C–C=C).

Methylene-interrupted—Describing a situation in which double bonds between carbon atoms occur in a series with two single bonds in between (C=C–C–C=C).

Box 1-3. Saturation

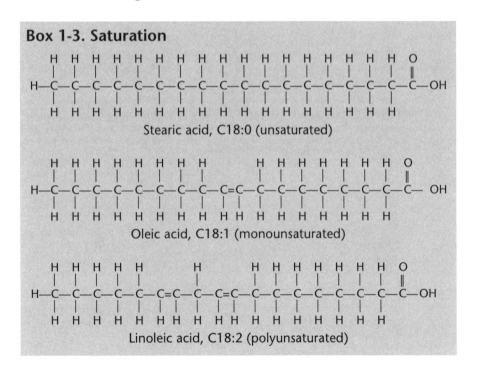

Stearic acid, C18:0 (unsaturated)

Oleic acid, C18:1 (monounsaturated)

Linoleic acid, C18:2 (polyunsaturated)

Box 1-4. An Omega Fatty Acid

There are six carbons between the terminal methyl group and the double bond, so linoleic acid is designated an ω6 fatty acid.

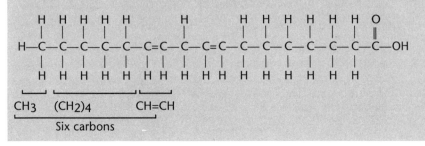

Box 1-5. Positions of Double Bonds

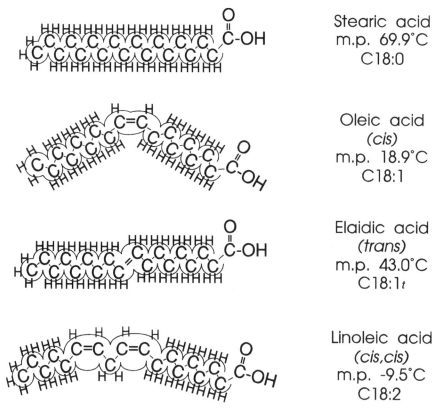

Conjugated   Methylene-interrupted   Isolated

two sides of the double bond are bent toward each other, and the hydrogen atoms on the double bond are on the same side (see oleic acid in Fig. 1-1). In the *trans* configuration, the hydrogen atoms on the double bond are opposite each other. As a result the chain is nearly straight (with a slight kink at the double bond), as shown for elaidic acid in Figure 1-1. The *cis* isomers prevail in all the food fats and oils, although small amounts of *trans* isomers occur in fats from ruminants.

Stearic acid
m.p. 69.9°C
C18:0

Oleic acid
(*cis*)
m.p. 18.9°C
C18:1

Elaidic acid
(*trans*)
m.p. 43.0°C
C18:1*t*

Linoleic acid
(*cis,cis*)
m.p. -9.5°C
C18:2

Fig. 1-1. Structure and melting point (mp) of several fatty acids. The types, from top to bottom are a saturated acid (stearic, C18:0), a *cis* monounsaturated acid (oleic, C18:1), a *trans* monounsaturated (elaidic, C18:1t), and a *cis, cis* polyunsaturated acid (linoleic, C18:2). Note that the melting point is higher for the saturated fatty acid than for the unsaturated ones and that the fatty acid with a *trans* double bond has a higher melting point than the fatty acid with the *cis* double bond.

The melting point of fatty acids varies according to some simple rules:

- Increasing the chain length increases the melting point.
- Increasing the saturation increases the melting point.
- Changing a *cis* to a *trans* isomer increases the melting point.

These relationships can be seen in Figure 1-1. (However, the first point is not strictly true; the odd-numbered fatty acids [2n+1] melt about 3°C lower than their even-numbered predecessors [2n]. For instance, C12:0 melts at 44.2°C , C13:0 at 41.4°C, and C14:0 at 54.4°C. Nevertheless, since most natural fatty acids are even-numbered, the rule can be taken as general.) The difference in melting points relates to how well the individual aliphatic chains pack in the crystal state; the closer and more uniform the packing, the higher the melting point.

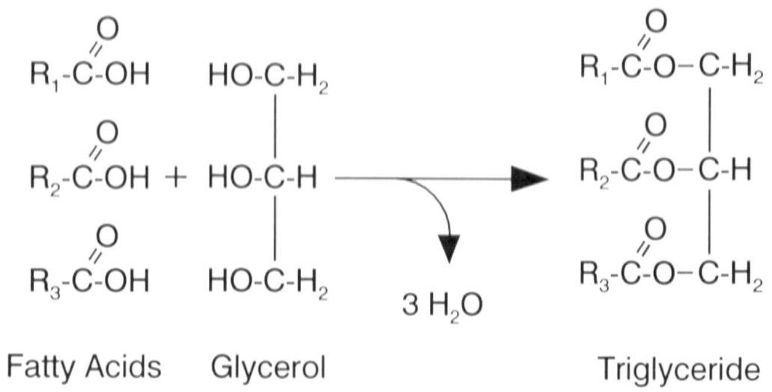

The same chain packing concepts apply to fatty acid chains when they are part of a triglyceride, and so the influence of fatty acid characteristics on the melting point of fat follows the same general rules. The conversion of a (liquid) vegetable oil to a (semi-solid) fat by hydrogenation (see Chapter 4) involves two changes: a decrease in degree of unsaturation and the isomerization of some *cis* double bonds to the *trans* configuration. Both changes increase the melting point.

Glycerides. As stated above, fats are esters, most commonly derived from the reaction of a single molecule of the alcohol glycerol and three molecules of fatty acids to yield one molecule of a triglyceride and three molecules of water (Fig. 1-2). When the fatty acids are identical, the product is a simple triglyceride. An example is triolein, a major triglyceride in olive oil, in which all three fatty acids are oleic acid. A mixed triglyceride has two or three different fatty acids joined to the glycerol. An example is palmitooleostearin (usually abbreviated POS), a component of cocoa butter, in which R_1 is palmitic acid, R_2 is oleic acid, and R_3 is stearic acid.

Some connections between triglyceride melting point and fatty acid composition are displayed in Table 1-1. The three groups show three different relationships.

- The first group shows the effect of varying chain length, unsaturation, and cis-trans isomerization in a simple triglyceride.

Fatty Acids   Glycerol   Triglyceride

Fig. 1-2. Structure and formation of triglycerides.

Glycerides—Compounds that have one or more fatty acids attached to glycerol.

TABLE 1-1. Melting Points for Some Pure Triglycerides

Triglyceride	Abbreviation[a]	Melting Point (°C)
Group A[b]		
Tripalmitin	PPP	66
Tristearin	SSS	73
Trielaidin	EEE	42
Triolein	OOO	5
Trilinolein	LLL	-13
Group B		
Palmitooleopalmitin	POP	37
Palmitooleostearin	POS	37
Stearooleostearin	SOS	43
Group C		
Palmitooleopalmitin	POP	37
Palmitoelaidopalmitin	PEP	55
Stearooleostearin	SOS	43
Stearoelaidostearin	SES	61

[a] P = palmitic acid (C16:0), S = stearic acid (C18:0), O = oleic acid (C18:1, *cis*), E = elaidic acid (C18:1, *trans*), L = linoleic acid (C18:2, *cis, cis*). Note how saturation and/ or *trans* isomers in the component fatty acids raise the melting point of the triglyceride.

[b] Group A contains simple (monoacid) triglycerides. Examples in Group B represent 85% of the composition of cocoa butter. Group C shows mixed triglycerides.

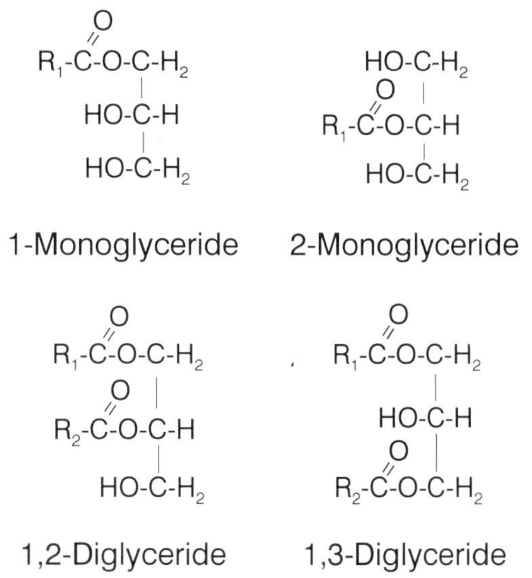

O
‖
R₁-C-O-C-H₂
|
HO-C-H
|
HO-C-H₂

1-Monoglyceride

HO-C-H₂
O |
R₁-C-O-C-H
‖ |
HO-C-H₂

2-Monoglyceride

O
‖
R₁-C-O-C-H₂
O |
R₂-C-O-C-H
‖ |
HO-C-H₂

1,2-Diglyceride

O
‖
R₁-C-O-C-H₂
|
HO-C-H
O |
R₂-C-O-C-H₂
‖

1,3-Diglyceride

Fig. 1-3. Structures of monoglycerides and diglycerides.

Diglyceride—A compound with a glycerol molecule attached to two fatty acids.

Monoglyceride—A compound with a glycerol molecule attached to one fatty acid.

Melting point—The temperature at which a solid turns into a liquid. Because they are a mixture of compounds, fats appear to melt over a range of temperature. A specific melting temperature is determined by warming a fat and recording the temperature at which an observable event coinciding with conversion to a liquid occurs.

- The second group shows the melting point of the three mixed triglycerides that together make up about 85% of cocoa butter.
- The third group shows the effect of cis-trans isomerization in mixed triglycerides.

If one or two fatty acids are removed (hydrolyzed) from a triglyceride, a *diglyceride* or a *monoglyceride* is formed. There are two positional forms of each compound, as shown in Figure 1-3, where the hydroxyl locations on glycerol are numbered from the top down. A monoglyceride can be 1-monoglyceride or 2-monoglyceride. (Note that 3-monoglyceride is the same as 1-monoglyceride). The two forms of a diglyceride are 1,2- and 1-3-. (Again, 2,3-diglyceride is the same as 1,2-diglyceride.) When hydrolysis is done with a simple catalyst such as sodium hydroxide, the monoglyceride. formed is mostly 1-monoglyceride, with 5–8% 2-monoglyceride. The secondary ester at the 2 position of glycerol is less stable, and migration of the fatty acid to the 1 position occurs. When the enzyme lipase is used to catalyze the hydrolysis, however, the result is different. This enzyme is specific for esters in the 1- and 3-position, and the product of the enzymatic reaction is only 2-monoglyceride, with 1,2-diglyceride being the only intermediate product.

Commercial plastic monoglycerides are made by adding glycerol (rather than water) plus a catalyst to the fat and heating to speed up the glycerolysis reaction. The reaction products are mainly 1-monoglyceride and 1,3-diglyceride, with approximately 5% triglyceride. The method for assaying monoglyceride in these materials measures only 1-monoglyceride, so a "50% -monoglyceride" product actually contains about 53% total monoglycerides. The production and characteristics of monoglyceride emulsifiers are discussed at more length in Chapter 3.

FUNCTIONAL CHARACTERISTICS

Solubility. Fat (triglyceride) is mostly aliphatic hydrocarbon chains and therefore is soluble in hydrophilic solvents such as hexane, benzene, and acetone and insoluble in water. Natural fats (and oils) contain small amounts of more hydrophilic materials, e.g., free fatty acids, monoglycerides, phospholipids, glycolipids, and oxidation products. These are more or less water-soluble, and are removed by washing with water during refining.

Melting points. When a pure chemical compound is heated, it undergoes a phase transition from solid to liquid at a sharply defined temperature. When the liquid is cooled, the transition is reversed at the same temperature. The sharpness of the melting point is one test of the purity of a material. Natural fats, however, are not pure compounds but rather a mixture of triglycerides made up of a variety of fatty acids. Thus, fat melting is gradual, and the definition of the melting point is dependent upon the method being used. The various methods are discussed in more detail in Chapter 2. When the melting point is included as one of the specifications for a fat, the method used must be stated, so that the vendor and buyer are using the same language.

Solidification and crystal structure. When a melted fat cools, the transition from liquid to solid is poorly defined and depends in part upon the rate of cooling. If the rate is very rapid (for instance, if melted fat is poured onto a block of ice), the fat solidifies into a waxy material (resembling paraffin wax) that is termed α *crystals*. If cooling is extremely slow, the highest-melting triglycerides in the fat have time to form stable β crystals. With intermediate cooling rates, the fat first forms α crystals, which rather quickly melt and reform into the metastable β′ crystals. The difference between the three crystal types has to do with the arrangement (crystal packing) of the fatty acid chains. The melting point of the crystal forms is in the order α < β′ < β. For pure tristearin (glycerol tristearate), they are 54.7, 63.2, and 73.5 °C, respectively.

In a crystal, the triglyceride is shaped like an elongated "h" (Fig. 1-4a). These are stacked in pairs, and the pairs then coalesce sideways to form layers in the crystal. The difference between the three crystal structures is mainly in the relative orientation of the pairs, viewed endways (Fig. 1-4b). In the α crystals, the pairs are oriented almost randomly; in β′ crystals, alternate rows are at right angles; in β crystals, the rows are all parallel. α Crystals are rather random in shape and dimensions, while β′ crystals are needle-shaped and about 5 μm maximum in length. β Crystals are blocky, about 50–100 μm on a side. These characteristics are obvious in photomicrographs of the crystals (Fig. 1-5). In a relatively pure triglyceride, the β′ crystal transforms to the β form fairly quickly, while in mixed triglycerides (the usual situation with natural fats), this change takes much longer. The fatty acids, which vary in chain length and degree of unsaturation (shape), require time to rearrange into the dense three-dimensional packing characteristic of β crystals.

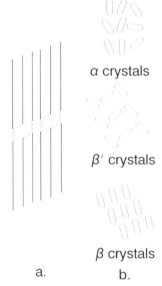

α crystals

β′ crystals

β crystals

a.   b.

Fig. 1-4. Molecular orientation in triglyceride crystals. a, The molecule is shaped like an elongated *h*. The wavy portion in the middle represents the glycerol molecule, and the linear parts represent fatty acid chains. b, Relative orientation of the molecular pairs in the crystal, viewed from the end. From top to bottom: α-crystals, β′-crystals, and β-crystals.

Crystals, α, β, β′—When triglyceride molecules in a fat turn from a liquid to a solid as a result of decreasing temperature, they pack into one of three different types of arrangement. Crystal forms exist only when the fat is in the solid state. They can affect the physical properties and functionality of the fat.

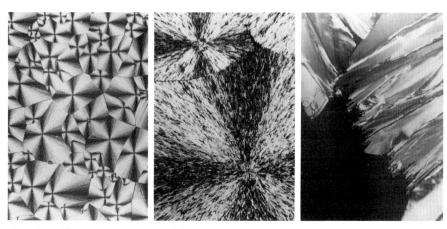

Fig. 1-5. Photomicrographs of fat crystals in polarized light. From left to right, α-crystals, β′-crystals, and β-crystals.

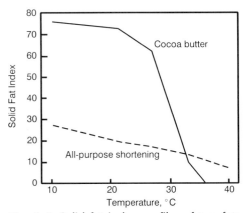

Fig. 1-6. Solid fat index profiles of two fats, cocoa butter and all-purpose shortening.

Solid fat index—A measure of the amount of solid fat in a fat at various temperatures. It is determined by the volume changes that occur as a result of melting or crystallization. This index relates the proportion of liquid to solid fractions in a fat.

Solid fat content—A measure of the amount of solid fat in a fat at various temperatures, determined by nuclear magnetic resonance. It is considered a more direct measure than the solid fat index.

A fat that has the β' crystal form is smooth and creamy. A fat that has transformed to β crystals, on the other hand, has a brittle, sandy texture. For most uses, both commercial and in the home, the former characteristics are preferred. Manufacturers of plastic shortenings and margarines use the β' stability of mixed triglyceride fats to produce products with the desired creaminess (see Chapter 4).

Solid to liquid ratios. A natural fat is a mixture of triglycerides, some solid and some liquid at any given temperature. The ratio of solid to liquid phase is an important determinant of fat functionality. It is expressed as the *solid fat index* (SFI) or the *solid fat content* (SFC; the differences between these two terms are discussed in Chapter 2). The SFI profiles for two fats are shown in Figure 1-6. All-purpose shortening (e.g., Crisco) has a relatively flat profile. It is soft enough to be worked at 10°C, yet still retains some solidity at 37°C (body temperature). By contrast, cocoa butter has a high profile with a rather steep descent. It is quite hard at 25°C, yet is liquid at 35°C. As the name implies, a high SFI value for any given temperature indicates that the fat is hard at that temperature. Roughly speaking, at an SFI greater than 35 (for example, butter at refrigerator temperature), the fat is hard and not readily spread, while at an SFI below 10 it is soft and almost liquid. The relationship between SFI and functionality in various applications is explored in depth in later chapters.

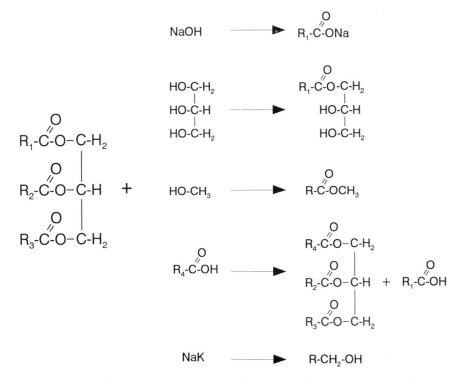

Fig. 1-7. Ester-splitting reactions: a, hydrolysis (saponification); b, glycerolysis; c, methanolysis; d, interesterification; and e, reduction.

CHEMICAL REACTIONS

Ester cleavage. The ester bonds in fat can undergo a variety of splitting reactions (Fig. 1-7). Some of these are important in food applications, while others have other industrial significance.

Hydrolysis. The addition of water yields a free fatty acid and a free hydroxyl group. This reaction, called *saponification*, is usually catalyzed with a base such as sodium hydroxide, and the fatty acid is neutralized to a sodium soap.

Glycerolysis. Glycerol can be the hydroxyl donor, forming a monoglyceride with the fatty acid and leaving a mono- or diglyceride behind. This reaction is the basis for making monoglyceride emulsifiers for food use. In commercial production, a basic catalyst, usually potassium carbonate, is used.

Alcoholysis. The reaction of fat with alcohol is usually catalyzed with an acid such as HCl or a sulfonic acid resin. For example, reaction of fat with methanol yields the methyl esters of the fatty acids, which are used to analyze the composition of the fat by gas-phase chromatography.

Interesterification. A free fatty acid can displace another fatty acid from an ester, leaving a glyceride with somewhat changed properties because its fatty acid structure has changed. This reaction is used to change fat properties. Lard, for example, has a nonrandom fatty acid distribution in its triglycerides. All the palmitic acid (25% of the total) is found in the 2 position. As a result, lard crystallizes rather readily in the β form, which is not desirable for bakery purposes. Heating in the presence of sodium methoxide or metallic sodium causes the fatty acids to shuffle their positions in the triglycerides randomly. The resulting fat has a lower SFI profile and is stable in the β′ crystal state, giving a more plastic shortening. Another application involves blending 20 parts fully hydrogenated soybean oil (which forms β crystals) with 80 parts refined soybean oil, then interesterifying. The product has the SFI profile appropriate for soft tub margarine, is β′ stable, and has only a trace of *trans* double bonds.

Reduction. Fat ester bonds can also be split reductively, yielding glycerol and long-chain alcohols. The reductant is an amalgam of sodium and potassium metals. The resulting fatty alcohols are used to make various detergents and lubricants (waxes) and are of great economic importance.

Oxidation. *Autoxidation* of fats occurs with unsaturated fatty acid chains. The relative rates of oxidation of oleic, linoleic, linolenic, and arachidonic acids (which have one, two, three, and four double bonds, respectively) are 1, 12, 25, and 50. The double bonds in the polyunsaturated acids are separated by methylene groups and are *cis* in their configuration.

Autoxidation is a series of *free radical* reactions, initiated and propagated by free radicals reacting with methylene $-CH_2-$ groups that are adjacent to double bonds (Fig. 1-8). (A free radical is an unpaired electron, indicated as a heavy dot in chemical formulas. It is a very reactive species.) At the beginning of the autoxidation process, a

Hydrolysis—A chemical reaction in which a molecule splits into two parts. A molecule of water also splits into H and OH, which are added to the places where the original bond was broken. A fatty acid is removed from a glyceride by hydrolysis of the ester bond.

Saponification—A chemical reaction caused by addition of alkali in which the fatty acids attached to a glycerol are cleaved off to produce soap (fatty acid salts) and a glycerol molecule.

Glycerolysis—A chemical reaction in which glycerol is combined with one or more fatty acids to form a glyceride.

Alcoholysis—A chemical reaction in which fatty acids react with alcohol to form an ester.

Interesterification—Changing the positions of the fatty acids on triglycerides. This is a commercial processing step to change the physical properties of a fat.

Reduction—Changing an acid group on a fatty acid to an alcohol group. This is done with metal reducing agents to create fatty alcohols for industrial uses.

Autoxidation—A reaction in which fats undergo oxidative changes due to the double bonds in their structure. The reaction can initiate and proceed without outside influences.

Free radical—An unpaired electron that is an unstable intermediate in the development of lipid oxidation and rancidity.

Peroxyl radical—An intermediate in lipid oxidation in which the fatty acid radical has added two oxygen atoms and is still a free radical. It is characterized by the structure COO•.

Hydroperoxide—An intermediate in lipid oxidation in which the fatty acid has added two oxygen atoms and a hydrogen atom at the point of oxidation. It is no longer a free radical but eventually degrades to flavor compounds associated with rancidity.

Rancidity—An off flavor in a fat or oil caused either by oxidation or by the release of flavorful fatty acids from the triglyceride.

Stability—The resistance of a fat source to the formation of rancidity.

Antioxidants—Compounds that can inhibit the development of lipid oxidation.

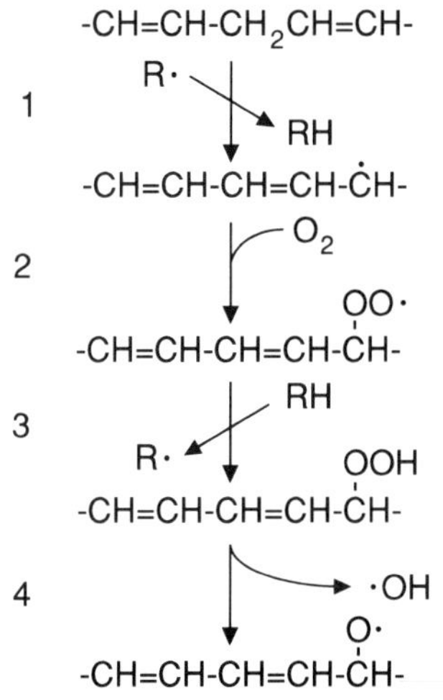

Fig. 1-8. Reactions occurring during autoxidation of fat.

hydrogen radical is extracted, and one of the double bonds shifts, moving the radical site to the outer carbon (reaction 1). Dissolved oxygen adds to this site, generating a *peroxyl radical* (reaction 2); this abstracts a hydrogen from a donor—perhaps another methylene group—making a *hydroperoxide* (reaction 3). The hydroperoxide splits to generate two free radicals, a hydroxyl and an alkoxyl radical (reaction 4). This cleavage is catalyzed by traces of metal ion such as copper or iron. The net result is three free radicals, each of which can initiate another chain of reactions. The rate of reaction is self-enhancing, i.e., it is an *autocatalytic* reaction.

The signs of *rancidity* (musty odors; bitter, disagreeable flavors) are due to breakdown products of the alkoxyl radical structure. These products are a variety of aldehydes and ketones derived from breaking the fatty acid carbon chain at the point where it is oxidized. Common products are heptanal, ethyl hexyl ketone, and the ω-aldehyde of nonanoic acid.

The reactions described above can occur in the dark, as long as molecular oxygen and an initiating free radical species are present. If the oil is exposed to light, oxygen may be photoactivated to singlet oxygen, which can initiate the chain at the second step shown in Figure 1-8.

In summary, four main factors contribute to autocatalytic rancidity:

- Chain initiation by trace free radicals
- Chain propagation by molecular oxygen
- Hydroperoxide cleavage catalyzed by metal ions
- Chain initiation by photoactivated oxygen.

These factors can be minimized by good manufacturing practices. The trace free radicals arise from peroxides that are left behind from inadequate refining and deodorization. Molecular oxygen should be excluded by processing, transporting, and storing oil under a nitrogen atmosphere. Metal ions can be kept out of the oil by having properly designed and maintained equipment, and traces of metal in the oil can be inactivated by chelation with citric acid. Finally, the exposure of oil to light should be minimal. With these precautions, oil oxidative *stability* can be increased several fold.

Oxidative stability can also be increased with *antioxidants*. These react with the active free radicals, transferring the radical function to the antioxidant (Fig. 1-9). Because of the ring structure of the antioxidant, this radical has low reactivity and does not initiate new reaction chains. However, if free radicals continue to form, due to the presence of oxygen and trace metals, eventually all the antioxidant will react, and the autocatalytic sequence will develop without hindrance.

Several organic compounds are in use today as antioxidants. They all have in common the ring structure shown in Figure 1-9, but they vary somewhat in the structure of the side groups. The ones approved by the Food and Drug Administration (FDA) for use in foods are:

- BHA, butylated hydroxyanisole (shown in Fig. 1-9)
- BHT, butylated hydroxytoluene (as in Fig. 1-9, with -CH3

replacing -OCH3)

- TBHQ, tertiary butylhydroquinone (as in Fig. 1-9, with -OH replacing -OCH3)

- PG, propyl gallate (the n-propyl ester of 3,4,5-trihydroxy-benzoic acid).

These may be added to a fat or oil at a maximum level of 0.02% (singly or in combination). They are also allowed in other foods, but the allowable maximum usage level varies with the food. The supplier of antioxidant is the best source of information with respect to particular food products.

FAT AND OIL SOURCES AND COMPOSITIONS

Fatty acid composition. The fatty acid composition of fats and oils has a great deal to do with their functionality. This relates not only to direct use in various foodstuffs, but also to the amenity of the fat to various processing steps such as hydrogenation, to its stability under storage, and to its nutritional significance. A compositional breakdown of edible fats and oils is given in Appendix B.

Lauric fats. These receive their name from the high content of lauric (C12:0) acid. They typically have a rather steep SFI profile. Partial hydrogenation of the relatively small amounts of oleic and linoleic acids moves the SFI curve laterally (to higher temperatures) and raises the melting point. Lightly hydrogenated coconut or palm kernel fat was traditionally used to make the filling creme for sandwich cookies but now has been largely replaced with selectively hydrogenated soy or cottonseed oil.

Vegetable fats. These can be divided into three groups:

- Saturated: cocoa butter, palm oil

- Oleics: olive oil, peanut oil, canola, high-oleic safflower, high-oleic sunflower

- Linoleics: corn, cottonseed, safflower, soybean, sunflower.

The saturated oils are used as is or partially hydrogenated (e.g., palm oil) to raise the SFI profile. The oleic oils are also used as is, primarily as salad oil or for light-duty frying. The linoleic oils may be used as is for salad oil but are usually hydrogenated to some degree. Light hydrogenation gives a salad and cooking oil with increased oxidative stability. Partial hydrogenation under various conditions gives fats with varied SFI profiles (steep, shallow, humped) for different applications. This ability to generate a "tailored" SFI profile is unique to the linoleics.

Animal fats. Butter, lard, and *tallow* were the traditional fats of northern Europe. Their availability depends in part upon a vigorous animal husbandry industry. As the per capita consumption of beef and pork

Lauric fats—A group of fat sources that are high in lauric acid as a component of the triglycerides.

Vegetable fats—Fats and oils derived from plant sources.

Animal fats—Fats (like butter, lard, and tallow) derived from animals.

Tallow—A hard white fat obtained from beef or sheep.

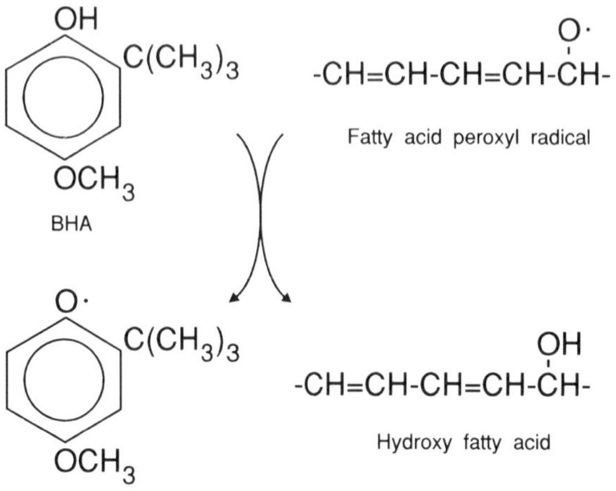

Fig. 1-9. Termination of the chain of autoxidation reactions by the antioxidant butylated hydroxyanisole (BHA).

Cholesterol—A fat soluble compound found in animal products that is required by humans, is produced by the body, and, if present at high levels in the blood stream, is associated with increased risk of diseases of the circulatory system.

has declined in the United States, the amount of edible tallow and lard has also declined. These fats all carry some amount of *cholesterol* with them, and because of the connection between dietary cholesterol and cardiovascular disease, their consumption has declined markedly.

Fish oil. Historically, marine oils have been an important fat source, but their use has declined since midcentury. Menhaden oil, in the partially and fully hydrogenated forms, has now been approved by the FDA for food. Marine oils are rich in long-chain polyunsaturated fatty acids (C20 and up). In particular, they contain large amounts of ω-3 fatty acids (i.e., acids in which the terminal double bond is three carbon atoms removed from the end of the fatty acid chain). There is some evidence that these fatty acids have a positive nutritional impact, probably as precursors to various prostaglandins.

Other components

Phospholipids—Natural components of fat that have phosphorous associated with the glycerides. Phospholipids are surfactants that assist in emulsification.

Phospholipids. Vegetable oils contain 0.1–3% phospholipids, which are removed during refining. Collectively called *lecithins*, they comprise several different chemical forms. Details of lecithin structure are discussed in Chapter 3.

Waxes. Waxes are found in small amounts in most unrefined oils. These are esters composed of a fatty acid linked to a fatty alcohol. A typical wax, for example, may be made of stearic acid esterified to stearoyl alcohol. While present in quite small amounts, waxes solidify when chilled and must be removed during refining.

Lecithins—A phospholipid found in egg yolk and soybeans and also used as a food ingredient. It is a surfactant that can stabilize emulsions.

Sterols. Fats and oils contain a few tenths of a percent of sterol. In animal fats this is mainly cholesterol, while in vegetable oils it is a mixture called phytosterol.

Hydrocarbons. Most fats contain small amounts (less than 1%) of hydrocarbons. The most common is squalene, a highly unsaturated compound closely related to carotenes.

Waxes—Lipid compounds that are fatty acids linked to long-chain fatty alcohols. They occur naturally in unrefined oils.

Sterols—Lipid compounds found in trace amounts that have ring structures rather than the straight chains associated with fatty acids. Examples are cholesterol in animal products and phytosterols in plant products.

Hydrocarbons—Lipid compounds found in trace amounts in fats and oils. They are unsaturated carbon chains such as the compound squalene.

Analytical Tests of Fats and Oils

Analytical tests are important at every stage in the use of fats and oils. Measurements made on crude oil help determine some of the refining parameters, while measurements on fat and oil products ensure that individual refining steps have been carried out correctly. The finished product must match the expectations of the customer for that product. The fat processor's finished product specifications are (or should be) the same as the customer's raw material specification. To make these match, both parties (vendor and buyer) should be using the same tests, performed by qualified analytical personnel.

There are two manuals of standard analytical methods for fats and oils. At least one, if not both, of these must be on the shelf of the quality control laboratory. These are:

- D. Firestone, ed. *Official Methods and Recommended Practices of the American Oil Chemists Society*, 4th ed. AOCS, Champaign IL, 1990.

- C. Paquot and A. Hautfenne, eds. *IUPAC Standard Methods for the Analysis of Oils, Fats and Derivatives*, 7th ed. Blackwell Scientific Publications, Oxford, London, UK, 1987. A. Dieffenbacher and W. D. Pocklington, eds., 1st Supplement, Blackwell, 1992.

Procedures for performing most of the tests discussed below are given in one or both of these sources. In some instances, no standard method is available, and a pragmatic alternative is suggested. The specific method references for the various tests are given in Table 2-1.

Physical Tests

SOLID FAT INDEX/CONTENT

The solid fat index (SFI) and solid fat content (SFC) relate to the percent of shortening that is solid at various temperatures. This curve can have a variety of shapes; for cocoa butter it is humped, but for some others it is almost straight over most of the range, with a steeper or shallower slope. The whole curve cannot be predicted from a determination made at just one temperature; the entire SFI (or SFC) curve is required in order to understand the properties of the shortening at different temperatures. Curves for different fats may cross, so it is

In This Chapter:
Physical Tests
Solid Fat Index/Content
Melting Point
Cloud Point
Oxidative Stability
Fire Danger Points
Polar Lipids in Frying Fat
Sensory
Plasticity
Chemical Tests
Fatty Acids
Iodine Value
Oxidative Degradation
Phosphorus (Residual Gums)
Unsaponifiable Material
Product Specifications

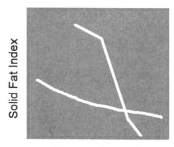

Examples of SFI curves.

Plastic fat—Fat that contains both solid and liquid triglycerides and at room temperature has a consistency that will hold its shape but is soft and pliable.

Basestocks—Fats with certain composition and melting characteristics that are mixed in order to get desirable melting properties in a margarine or shortening.

incorrect to compare the SFI/C of two fats without giving the temperature at which the comparison was made.

The functionality of a *plastic fat* depends not only on the solids content, but also on the slope of the SFI/C curve. Meeting an SFI/C specification implies that the values for a particular batch lie within the specified ranges and also that deviations from the target values are all on the same side, either higher or lower. This is particularly important in producing *basestocks* for blending to produce margarine or shortening.

The original method for estimating the percent of solids in a fat was based on *dilatometry* (see Box 2-1: Measurement of Solid Fat Index, and Fig. 2-1); the numbers obtained are referred to as the "solid fat index." This method is time-consuming and subject to bias. More recently low-resolution nuclear magnetic resonance (NMR) has been used to measure the relative amounts of liquid and solid fat in a sample, based upon the difference in rates of relaxation of protons in the two phases after the sample has been pulsed. With proper calibration this gives a direct determination of the percentage of solid fat, and the results are termed "solid fat content." The analysis takes less time than dilatometry, but the equipment is more expensive.

The relationship between SFI and SFC is a function of temperature, type of shortening, and the level of SFI. Comprehensive studies of 46 shortenings across the temperature range of 10–45°C and of 13 confectionery fats *(hard butters)* provided the data for the plots shown in Figures 2-2 and 2-3. For shortenings, the plots are different at different temperatures, while for hard butters, the relationship is essentially the same at all temperatures. The authors of these figures (1,2) give tables for interconverting SFI and SFC.

Analytical measurements of SFI and SFC are relatively precise; duplicate determinations should agree within ±1 unit. The suggested values given in Appendix C for SFI and SFC have rather large tolerances specified, because it is difficult to control the steps in shortening production to closer tolerances than those given. In any event, it is important that the specification be clear as to which method is to be used: dilatometry (for SFI) or NMR (for SFC).

TABLE 2-1. Methods for Analysis of Edible Fats and Oils

Test	AOCS	IUPAC
Solid fat index (dilatometry)	Cd 10-57	2.141
Solid fat content (pNMR)	Cd 16b-93	2.150
Melting point		
Complete mp	Cc 1-25	…
Wiley mp	Cc 2-38	…
Dropping point	Cc 18-80	…
Slip point	Cc 3-25	…
	Cc 3b-92	
Cloud point	Cc 6-25	…
Oxidation		
Active oxygen method	Cd 12-57	2.506
Oil stability Index	Cd 12b-92	…
Oxygen bomb	…	…
Smoke, flash, fire points	Cc 9a-48	…
Polar lipids in frying fat	Cd 20-91	2.507
Color		
Lovibond	Cc 13c-92	…
Spectrophotometric	Cc 13c-50	2.103
Plasticity (hardness)	Cc 16-60	…
Fatty acids		
Composition	Ce 1-62	2.302
Polyunsaturated	Cd 15-78	2.206
trans Double bonds	Cd 14b-93	2.207
Free fatty acids	Ca 5a-40	2.201
Iodine value	Cd 1-25	2.205
Oxidation		
Peroxide value	Cd 8b-90	2.501
Anisidine value	Cd 18-90	2.504
Totox number	…	…
Thiobarbituric acid	Cd 19-90	2.531
Phosphorus	Ca 12-55	2.421
Unsaponifiable material	Ca 6b-53	2.401

MELTING POINT

Melting point (mp) usually refers to the temperature at which a pure compound changes from the solid to the liquid phase. In the case of commercial fats, which are mixtures of triglycerides (and perhaps monoglycerides and other emulsifiers), no sharp change occurs. In fact the "melting point" of a fat is more accurately termed a "dissolution point." To illustrate, fully hydrogenated palm oil (consisting of mixed triglycerides of palmitic and stearic acids) melts at 58–60°C (135–140°F) and is added to plastic shortening at 5–10% of the total. The melting point of the resulting fat is usually around 45–47°C (113–117°F); the solid fat has fully dissolved in the oil phase at this temperature.

Complete melting point. This measurement is also sometimes referred to as a *capillary melting point*. The sample is introduced into a glass capillary tube, which is then sealed. It is then chilled to solidify the fat, then heated in a water bath. The temperature at which the last visible traces of solids disappear is the complete melting point. It corresponds to an SFI of 0.

Dilatometry—A technique for measuring the amount of solid or liquid in a fat based on small volume changes that occur when going from the solid to the liquid state.

Hard butters—Fats that are very firm at room temperature and have a high melting point.

Capillary melting point—The temperature at which a solid fat turns into a liquid, measured by warming a fat sample in a small tube and observing the temperature at which the solid character of the fat disappears.

Box 2-1. Measurement of Solid Fat Index

When a solid fat melts, it expands. In the case of a simple triglyceride (e.g., tripalmitin), this melting point, and the attendant increase in specific volume, occurs over a very narrow range. For a commercial fat, the change is more gradual because the fat is a mixture of many different triglycerides that melt over a range of temperatures. Solid fat index (SFI) is measured by placing a weighed sample of the fat in a dilatometer and measuring the volume at various temperatures (Fig. 2-1). At a given temperature, the specific volume (volume divided by weight) of the sample equals the specific volume of the solid phase times the fraction (X) of the sample that is solid, plus the specific volume of the liquid phase times the amount that is liquid:

$$\text{SpVol}_{\text{sample}} = [\text{SpVol}_{\text{solid}} \times X_{\text{solid}}] + [\text{SpVol}_{\text{liquid}} \times (1 - X)]$$

Ideally, X (times 100) would represent the SFI of the sample at the temperature of measurement.

However, there is an experimental difficulty in the exact determination of X by dilatometry. Like all materials, a triglyceride (either solid or liquid) expands when heated. (A solid fat has a coefficient of expansion of about $0.00038 \text{ ml} \cdot \text{g}^{-1} \cdot °\text{C}^{-1}$, while an oil has a coefficient of expansion of about $0.00084 \text{ ml} \cdot \text{g}^{-1} \cdot °\text{C}^{-1}$.) While the dependence of specific volume on temperature for a particular oil (the dashed line in Fig. 2-1) is easy to determine in the laboratory, the corresponding line for the fully solid fat is difficult to define. The standard SFI method circumvents this difficulty by adopting a convention: the lower reference line is given the same slope as the line for the liquid and is located 0.100 units (T) below it. (This is the dotted line in Figure 2-1.) Then X (the SFI) is defined as the difference (S) between the experimental specific volume ($\bullet$) and the extrapolated liquid specific volume, divided by T (that is, 0.1) and expressed as a percentage. In the example that Figure 2-1 illustrates, the SFI at 20°C is $S/T \times 100$, or 18.

This convention distorts the numbers, so that SFI is somewhat smaller than the actual percentage of solid fat in the sample. A hard fat (fully hydrogenated vegetable oil, melting point 60°C) has an SFI of about 80 at room temperature, even though it is essentially all solid. By extrapolation of the curves comparing SFI and solid fat content (SFC) in Figures 2-2 and 2-3, it is apparent that an SFC of 100 equates to an SFI around 75–80.

Wiley melting point—The temperature at which a solid fat turns into a liquid, measured by warming a fat until it loses its shape.

Wiley melting point. Fat is formed into a small disk (in a mold), then chilled. The disk is slowly warmed (in a water bath), and the melting point is the temperature at which the disk changes its configuration. The official method specifies that the fat should become spherical, but some analysts look for the point at which the sharp edges disappear and the fat takes on a football shape. The change is rather gradual, and disagreement between analysts is not unusual. For plastic fats, the Wiley melting point occurs when the SFI is about 2–3. For hard butters,

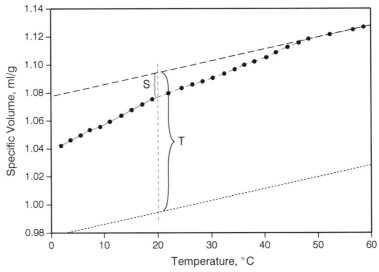

Fig. 2-1. Dilatometry of a commercial plastic shortening (Crisco). The dashed line shows the extrapolated volume of the liquid fat. The dotted line is the conventional line, placed 0.100 units (*T*) below the liquid line. The solid fat index (SFI) is the difference (*S*) between the actual measured volume (•) and the extrapolated liquid line, divided by *T* (which is 0.1) and expressed as a percentage. At 20°C, the sample shown had an SFI of 18.

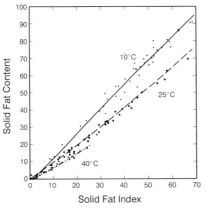

Fig. 2-2. Relationship between solid fat index (SFI) and solid fat content (SFC) for plastic shortenings. The data are shown for 10, 25, and 40°C. Tables for interconverting SFI and SFC are found in the original publication (1).

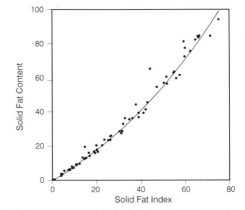

Fig. 2-3. Solid fat index (SFI) vs. solid fat content (SFC) for confectionery hard butters. Unlike that for plastic shortenings, the relationship for these fats shows almost no temperature dependence. Tables for interconverting SFI and SFC are found in the original publication (2).

the melting point is sharper, and the SFI may be somewhat less.

Dropping melting point. The Mettler Instrument Corporation has introduced an instrument that measures fat melting point automatically. The fat is placed in a cup that has a small orifice in the bottom. The cup is warmed at a constant rate. When the fat liquefies, it flows through the orifice, forming a drop that interrupts a light beam; this is the endpoint, which is automatically registered. No operator attention is required during this process, making the determination more objective than with the other methods. The dropping melting point corresponds to an SFI of about 1–2.

Slip point. This method is not commonly used in the United States but is frequently used in other countries. The sample is placed in a capillary tube, but the ends are left open rather than being sealed (as for the complete melting point). After chilling, the tube is placed vertically in a water bath and slowly warmed. The slip point is the temperature at which the fat rises a specific distance in the capillary. It corresponds to an SFI of about 5 (somewhat less for hard butters).

CLOUD POINT

In some applications, mayonnaise and salad dressings in particular, it is important that the oil remain completely liquid at refrigerator temperatures. To test this capability, the oil is held in an ice bath, and the time until the first cloudiness appears is noted. A time of 5.5 hr is considered minimum, and a *cloud point* time of 20 hr is quite good. For monitoring oil during manufacture, a rapid test is used. The oil sample is chilled at -60°C (-76°F) for 15 min, then held at 10°C (50°F). If no solid material is visible after 30 min, the oil passes the test.

OXIDATIVE STABILITY

Resistance to autoxidation (see Chapter 1) is important in fats and oils, more so in some applications (e.g., frying fat, cracker spray oil) than in others (e.g., bread shortening). To measure this resistance, the oil is heated in the presence of oxygen, and the length of time needed for the peroxide value (see below) to reach a specified level is determined.

Active oxygen method (AOM). Fat oxidative stability is measured by the AOM. Air is bubbled through oil or fat held at 97.8°C. Periodically, oil samples are withdrawn and the *peroxide value* is determined. The time required to develop a peroxide concentration of 100 meq/kg is the AOM stability of the sample. (This method is now considered obsolete. The oil stability index is now standard in the industry, although product specifications still routinely give AOM values.)

Oil stability index (OSI). The endpoint for oil stability is measured automatically in this test. Air is bubbled through hot oil, and one of the breakdown products (formic acid) is carried by the stream of air into distilled water in a cell. The machine continuously monitors conductivity of the water. The time at which it rises sharply is the end

Dropping melting point—The temperature at which a solid fat turns into a liquid, measured by warming the fat until it forms a drop of liquid.

Slip point—The temperature at which a solid fat becomes more liquid-like. It is measured by warming a fat sample in a small tube and observing the temperature at which it moves.

Cloud point—The amount of time an oil remains clear when cooled to refrigerator temperature.

Active oxygen method—A procedure to determine how rapidly an oil or fat oxidizes to form peroxides.

Peroxide value—A number that indicates the level of peroxides in a fat or oil that has developed as a result of oxidation. Peroxides are considered intermediates in the lipid oxidation reaction scheme.

Oil stability index—An automated procedure for determining the speed at which oxidized products develop in a heated oil when air is bubbled through.

point of the determination. The rate of the autoxidation reaction almost exactly doubles with each 10 degree C increase in temperature, and the apparatus is usually operated at 110°C to shorten testing time. The results correlate well with AOM values; results obtained at 110°C are 40–45% of the AOM numbers (determined at 97.8°C).

Oxygen bomb. The sample is sealed in a heavy-walled container attached to a pressure recorder. The bomb is pressurized with O_2 to 100 psi, then placed in a boiling water bath. The end point is the time when the sample starts to absorb oxygen rapidly and a sharp pressure drop occurs. It is reported that oxygen bomb results correlate better with product shelf life tests (appearance of rancidity) than do AOM numbers. This method has the added advantage that the whole sample (e.g., snack crackers) may be used in the test; no prior fat extraction is necessary.

FIRE DANGER POINTS

Smoke point, flash point, fire point. Fat is placed in a cup and heated in a white-walled chamber under a strong light. The temperature at which wisps of smoke are observed is the *smoke point*. With further heating, a small open flame that is passed over the hot oil causes flashes of burning fat; this is the *flash point*. Upon further heating, eventually a continuous fire can be ignited at the fat surface (*fire point*). The presence of free fatty acids, emulsifiers, and bits of fried food in the oil all serve to lower these temperature points (Fig. 2-4). The flash point is approximately 140 degrees C (250 degrees F) higher than the smoke point, and the fire point is 55 degrees C (100 degrees F) higher still. Smoke point is an important test for fats used in deep frying operations. If the fat has a relatively low smoke point to start with, the formation

Oxygen bomb—A procedure to measure the speed at which a heated sample consumes oxygen due to lipid oxidation.

Smoke point—The temperature at which a heated oil begins to give off smoke.

Flash point—The temperature at which a heated oil gives flashes of burning when exposed to a flame.

Fire point—The temperature at which a heated oil burns with a flame when ignited.

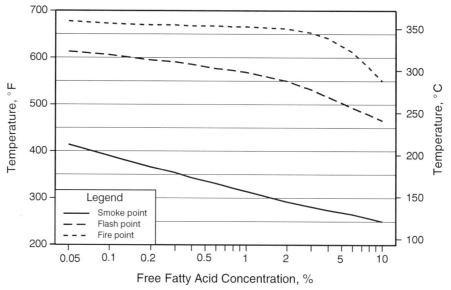

Fig. 2-4. Effect of free fatty acid content on the smoke, flash, and fire points of soybean oil.

of free fatty acids and fat breakdown products during use may lower the flash point sufficiently that the fat may be ignited by the heating elements.

POLAR LIPIDS IN FRYING FAT

The presence of oxidized fat and subsequent reaction products (ketones, aldehydes, polymerized materials) is a measure of the suitability of the fat for continued use in frying (see Chapter 6). The percentage of these polar materials in a sample of used fat is determined chromatographically. A weighed sample of fat is placed on the top of a silica gel column, and the *nonpolar lipids* (unchanged triglycerides) are eluted with solvent (petroleum ether/diethyl ether). The weight of the nonpolar lipids is determined after the solvent is removed by evaporation. The percentage of polar lipids is calculated by difference. In many countries, frying fat used in food preparation must be discarded if the polar lipid concentration is greater than 27%.

SENSORY

Color. Oils differ in their inherent depth of color, depending upon the source. However, if a refined oil from a specific source is darker than expected, it indicates that the oil was either improperly refined or was abused. Color is usually measured using the Lovibond Tintometer. Oil in a tube of standard dimensions is compared with colored glass standards, usually red and yellow. The analyst tries a combination of the glass standards until a match is obtained. The results are usually expressed as the numbers on the glass standards, i.e., 1.5R, 15Y. A spectrophotometric method for color evaluation is also available.

Flavor, odor. *Organoleptic evaluation* of oil is an empirical test, and there is no substitute for an experienced analyst. The main negative factors are reversion flavors and *oxidative rancidity*. Various oils, upon storage after refining, develop characteristic flavors (reversion flavors). Soy oil, for example, develops a "beany" note, while palm oil develops a "metallic" flavor. Odors due to oxidative rancidity are primarily caused by volatile aldehydic and ketonic compounds from the breakdown of fatty acid *peroxides*. Evaluation is best done using oil that has been warmed to about 40–50°C (104–113°F) in a water bath.

PLASTICITY

Relationship to SFI/C and crystal structure. Plasticity ("creaminess," "spreadability") of a shortening or margarine is an important property but one that is difficult to specify exactly, using analytical instruments. As a rough rule of thumb, a fat shows acceptable plasticity when the SFI is in the range of 10–25 units. It is also important that the crystal structure be stabilized in the β′ form; in this form, the small needles break readily under stress but also reform quickly when the stress is removed. Such a shortening, for example, functions properly as a roll-in fat in making layered doughs or in the creaming operation during cookie manufacture (see Chapter 5). By contrast, β crystals give the

Polar lipids—Fat components that are more like water and less like fat in their solubility properties. Introduction of oxygen or nitrogen atoms into lipid molecules makes them more polar.

Nonpolar lipids—Fat components that are like organic solvents and not like water in their solubility properties. Unaltered triglycerides are usually very nonpolar.

Organoleptic evaluation—Evaluating quality using a sense, such as taste or smell.

Oxidative rancidity—Off-flavors in a fat or oil due to the reaction of oxygen with fat molecules.

Peroxides—Oxidized fat molecules that eventually degrade to off-flavors.

Cone stress index—A method of measuring the softness or pliability of a fat.

shortening or margarine a hard, grainy texture.

Measurement. The method for measuring plasticity uses the ASTM grease cone penetrometer. The depth to which a standard cone penetrates the fat during 5 sec is measured. DeMan (3) has reviewed work with this machine. Using a "*cone stress index*," which is proportional to $1/(depth\ of\ penetration)^2$, he found a good relationship with "spreadability." This test may be used to write an objective specification for shortening hardness, for example, for roll-in fats on a Danish or puff pastry production line.

Chemical Tests

FATTY ACIDS

Composition. The fatty acid composition of fats and oils is becoming a routine test, in large part because of the requirements of the new food labelling regulations. The fatty acids are converted from glycerides to methyl esters, then analyzed by gas-liquid chromatography (GLC). The methodology in this area continues to evolve; the *cis*, *trans*, and geometric isomers of monounsaturated fatty acids, for example, can be separated and measured by GLC. Typical fatty acid compositions of edible fats and oils are presented in Appendix B.

Polyunsaturated fatty acids (PUFAs) are legally defined as containing *cis, cis* methylene-interrupted double bonds. They are measured as the oxidation products (conjugated diene) obtained upon reaction with the enzyme lipoxygenase. Geometric and *trans* isomers of PUFAs do not react with the enzyme.

The concentration of *trans* double bonds is measured by infrared spectrophotometry. The sharp absorption peak at 968 cm^{-1} is used to quantitate the amount of *trans* double bonds in the fat (in a solvent such as carbon disulfide). Because of the numerous combinations possible, it is difficult to separate by GLC all the various *cis*, *trans*, and geometric isomers of PUFAs found in partially hydrogenated vegetable oils.

Free fatty acids—Fatty acids with an acid group that is not chemically bound to an alcohol group. Usually fatty acids are bound to glycerol to form triglycerides and are therefore not free.

Free fatty acids. To measure free fatty acid (FFA) content, a sample of oil is dissolved in alcohol and titrated with alcoholic sodium hydroxide to a *phenolphthalein end point*. The results are given as percent free fatty acid (%FFA), calculated as oleic acid. Any acidic compound in the sample reacts with the hydroxide. Thus, citric acid, added to refined oil as a metal chelator, is titrated and increases the apparent FFA content. This possible complication must be kept in mind if a sample appears to have excessive FFA.

Phenolphthalein end point—A color indicator used to determine when all the acid groups have been reacted with basic groups from sodium hydroxide added during a titration.

IODINE VALUE

Iodine value—A test to measure the number of double bonds in a fat or oil. A higher value means more double bonds.

Chemical basis. Elemental iodine (I_2) can add to double bonds in unsaturated fatty acids. This reaction (with variations that speed up the test) is used to measure fat unsaturation. The results are expressed as grams of iodine absorbed by 100 g of fat. The reaction occurs with both *cis* and *trans* double bonds. Typical iodine values (IVs) are given for natural fats and oils in Appendix B.

Connection to physical properties. The conversion of a *cis* to a *trans* double bond raises the melting point of the triglyceride involved but does not affect IV. An oil may be hydrogenated to the same IV using, for example, two different catalysts and conditions, and the resulting products will have quite different SFI profiles, melting points, plasticity ranges, and other physical properties. Likewise a soy oil and a cottonseed oil at the same IV have quite different characteristics. Hence, IV correlates with physical functionality only when one compares the same starting oils and the same processing (hydrogenation) conditions.

OXIDATIVE DEGRADATION

Peroxide value (PV). Hydroperoxides formed by fat oxidation react with iodide ions to form iodine, which in turn is measured by titration with thiosulfate. The resulting PV is expressed as milliequivalents of iodine formed per kilogram of fat. This is the classical method for determining the extent of oxidation of a fat. As part of a raw material specification, it ensures that the ingredient has been properly refined and handled.

Anisidine value (AV). PV does not measure all the breakdown products formed from peroxides, in particular the α,β-unsaturated aldehydes. However, these form a colored adduct with anisidine; the color is quantitated at 350 nm and converted to AV. The test is seldom used in the United States but is widely used in Europe, particularly as a part of the Totox number.

Totox number. A measure of the total oxidation of the fat or oil is the Totox number:

$$Totox = AV + (2 \times PV) .$$

During storage stability testing of high-fat products, it is often observed that PV first rises, then falls (as hydroperoxides decay). Totox, however, measures both hydroperoxides and their breakdown products. It tends to rise continuously and gives a better measure of the progressive oxidative degradation of the fat.

2-Thiobarbituric acid (TBA) *value.* TBA forms colored adducts with the breakdown products formed from fatty acid peroxides. The aldehyde products form a yellow pigment with an absorption at 450 nm, while the 2,4-dienals (for example, from the breakdown of linoleic acid hydroperoxide) give a red adduct with absorption at 530 nm. The test is quite sensitive, registering the presence of fatty acid oxidative breakdown products at a very early stage in the development of rancidity.

PHOSPHORUS (RESIDUAL GUMS)

During the refining of corn and soybean oil, lecithins (phospholipids, gums) must be removed as completely as possible. Residual lecithin in the oil contributes to oxidative instability and leads to "fishy" taste and odor in foods fried in the oil. The standard method for measuring

Anisidine value—A test to determine the amount of reaction products produced by lipid oxidation.

Totox number—A measure of the total amount of intermediate compounds (peroxides) and end-product compounds that result from lipid oxidation.

2-Thiobarbituric acid value—A measure of lipid oxidation that determines the concentration of certain end products of lipid oxidation.

phospholipid content is to ash the sample and determine phosphorus in the residual ash colorimetrically or by atomic absorption spectrophotometry. A quicker method, useful for monitoring the completeness of degumming during refining, uses the fact that lecithin is insoluble in acetone (4). The oil sample is dissolved in acetone and made to a volume of 50 ml. The turbidity (due to insoluble gums) of the suspension is read in a nephelometer and is related to phosphorus content using a standard curve.

UNSAPONIFIABLE MATERIAL

Waxes, sterols, and hydrocarbons in fats and oils are generally determined as unsaponifiable material. The weighed sample is saponified (refluxed with aqueous sodium hydroxide). The cooled mixture is repeatedly extracted with petroleum ether and/or diethyl ether. The solvent is removed from the combined dried extracts, which are then weighed and reported as unsaponifiable material. The content of unsaponificable material in a properly refined fat or oil is generally less than 0.5% and is not a problem with respect to quality.

TABLE 2-2. Raw Material Specification for All-Purpose Shortening

Characteristic	Requirement	AOCS Method
Chemical		
Peroxide value	1 meq/kg maximum	Cd 8b-90
Free fatty acid (as oleic acid)	0.05% maximum	Ca 5a-40
Phosphorus content	1 ppm maximum	Ca 12-55
AOM stability	75 hr minimum	Cd 12-57
Oil stability index	30 hr minimum	Cd 12b-92
Wiley mp	46 ± 1°C	Cc 2-38
Physical		
Color (Lovibond)	1.5 R, 15 Y maximum	Cc 13c-92
Flavor	Bland	…
Odor	Neutral when warmed	…
Solid fat profiles	See chart below	SFC, Cd 16b-93
		SFI, Cd 10-57

°C	SFC	°F	SFI
10	38 ± 3	50	28 ± 3
15	29 ± 3	70	20 ± 2
20	22 ± 3	80	17 ± 1
25	17 ± 2	92	13 ± 1
30	12 ± 2	104	7 ± 1
35	8 ± 1		
40	4 ± 1		

Product Specifications

The main reason for expending resources to perform the tests discussed above is to help the buyer of a fat or oil communicate to the supplier precisely what is needed. This communication is in the form of a product (or raw material) *specification*. It is a written list of tests and the desired result for each analysis. It is the buyer's responsibility to issue this specification. Of course, frequently the actual numbers are determined through collaboration with the supplier(s). In essence, the buyer is saying "If you supply a material that meets these specifications, I will consider purchasing it from you."

A typical specification for an all-purpose shortening is presented in Table 2-2. This is only a partial specification. The buyer may want to specify the sources of the oils used to make the shortening, and the specification should also address such matters as packaging (kind, size), storage conditions, necessity for Good Manufacturing Practices at all stages of production, and other matters that are outside the specific topics covered in this chapter.

A brief comment on the reason for, and importance of, each of the items listed in Table 2-2 might be helpful.

- Peroxide value. Since peroxides are initiators of autoxidation (see Chapter 1), a PV greater than 1 meq/kg indicates that the fat will have reduced stability in the finished product.

- Free fatty acid. The free fatty acids in a fat are removed during refining and deodorizing (see Chapter 4). An FFA level of more than 0.05% indicates that these steps were not properly done, *or* that there is a high level of citric acid in the oil.

- Phosphorus content. More than 1 ppm phosphorus indicates incomplete degumming of the oil used to make the shortening. The residual lecithin promotes autoxidation (rancidity).

- AOM and OSI. One or the other of these is included, to ensure that the shortening has sufficient stability against oxidation for its intended purposes.

- Wiley melting point. This is unimportant for an all-purpose shortening but is significant for fats present in high concentration in the food product (e.g., sandwich cookie filler fats, confectionery coatings).

- Color (Lovibond). This primarily ensures that the oils used to make the shortening were properly refined and bleached. The exact numbers depend in part upon the parent oils.

- Flavor. If the shortening has any noticeable off-flavors, it should be immediately rejected. There are many sources for bad flavors, none of which are acceptable.

- Odor. The same comments apply here as for flavor. Any noticeable off-odor is grounds for immediate rejection of a shipment.

Specifications—A set of chemical and physical quality requirements that a product or ingredient must meet before it is acceptable.

- Solid fat profile. This is probably the most important part of a specification for a fat because it has the greatest influence on the suitability of the fat for a particular application. That influence is discussed in subsequent chapters on applications. It is important to know what method for solid fat measurement is being used by the supplier: pNMR (for SFC) or dilatometry (for SFI). In a general specification (for communicating to several suppliers), both kinds of profiles should be given.

A short summary of selected tests is given in Box 2-2.

There are other tests of importance to fat or oil for a particular use that are not listed here. They are given in the specific listing in Appendix C.

The method to be used for each of the tests specified should be listed. Then if there is ever a question about the analysis of a particular sample, all laboratories concerned (the supplier's, the buyer's, and any outside service laboratory, if needed) will be using the same methodology and should get the same result.

The specifications for many commonly used food shortenings and oils are given in appropriate chapters and summarized in Appendix C. These are representative of shortenings and margarines presently sold in the United States for industrial uses. They reflect average properties of products sold by several different suppliers. These specifications are

Box 2-2. Description of Selected Tests

Test/Determination	Description
Active oxygen method (AOM)	Measures peroxide content to show stability in oxygen
Anisidine value (AV)	Measures peroxides to indicate amount of fat oxidation
Cloud point	Time taken for chilled oil to become cloudy
Fire point	Temperature at which oil burns
Flash point	Temperature at which oil gives a flame
Iodine value (IV)	Measures double bonds to indicate degree of fat unsaturation
Melting point determination	
Capillary (complete)	Temperature at which fat in sealed tube melts
Dropping	Measured automatically by Mettler instrument as drop of fat falling through a light beam
Slip point	Temperature at which fat rises in a capillary tube
Wiley	Temperature at which a disk of fat loses its shape
Oil stability index (OSI)	Measures formic acid to show stability of fat in oxygen
Peroxide value (PV)	Measures hydroperoxides to indicate amount of fat oxidation
Plasticity	Amount of creaminess, measured by ASTM grease cone penetrometer
Smoke point	Temperature at which oil smokes
Solid fat content (SFC)	Amount of fat that is solid at a given temperature, measured by NMR
Solid fat index (SFI)	Amount of fat that is solid at a given temperature, measured by dilatometer
2-Thiobarbituric acid (TBA) value	Measures breakdown products of peroxides to indicate amount of fat oxidation
Totox number	Measures hydroperoxides and their breakdown products to indicate amount of fat oxidation

intended as suggested starting points for experimentation by users. From such tests, the buyer can develop specifications for products that give optimum performance with the equipment and conditions prevailing in a specific plant.

Certain characteristics are common to the specifications of all these products. To save space, these characteristics are listed once at the beginning of Appendix C. They should be included in each individual ingredient specification. Other specifications unique to each type of fat or oil are also given. The methods are not listed but can be readily filled in by reference to Table 2-1.

References

1. van den Enden, J. C., Haighton, A. J., van Putte, K., Vermaas, L. F., and Waddington, D. 1978. Fette, Seife, Anstrichmittel 80:180.
2. van den Enden, J. C., Rossell, J. B., Vermaas, L. F., and Waddington, D. 1982. J. Am. Oil Chem. Soc. 59:433.
3. DeMan, J. M. 1983. J. Am. Oil Chem. Soc. 60:82.
4. Sinram, R. D. 1986. J. Am. Oil Chem. Soc. 63:667.

Properties of Emulsifiers

Functions in Food Systems

Oil and water do not mix; oil is nonpolar, whereas water is a polar liquid. If oil and water are shaken together, an *emulsion* is formed. In a familiar example, when a bottle containing olive oil and vinegar is shaken, the oil phase separates into droplets suspended in the vinegar. The oil is called the "dispersed phase," while the vinegar is the "continuous phase" (Box 3-1). This example is an oil-in-water (O/W) emulsion. In other instances, for example, during the production of margarine, the water phase is dispersed in droplets in the oil phase, giving a water-in-oil (W/O) emulsion. If the mixture of oil and vinegar is left standing for a while, the two phases slowly separate, with the oil droplets rising to the top. The length of time for separation depends on the size of the droplets—large droplets rise faster than small ones. This process is called *creaming*. When the drops contact each other, they join together, or coalesce. The average drop size increases, and eventually a phase of oil becomes visible on the top. This is called emulsion "breakdown."

If an *emulsifier* is present in the system, the drops formed during shaking are smaller and take longer to rise to the top. Thus, a longer time elapses before the emulsion creams, the droplets in the dispersed phase coalesce, and the emulsion breaks down. Emulsifiers are sometimes called *surfactants*, but not all surfactants are effective emulsifiers. The key point is that good emulsifiers promote the subdivision of the dispersed phase. The phase in which the emulsifier is soluble is usually the continuous phase—a water-soluble emulsifier such as Polysorbate 60 promotes an O/W emulsion, while an oil-soluble emulsifier such as monoglyceride promotes a W/O emulsion. This tendency is expressed by the hydrophilic-lipophilic balance (HLB), which is discussed later in this chapter.

The actions of oil, water, and emulsifier in the situations described above can be explained in terms of how energy, both intrinsic and applied, influences the mixture of oil and water. Energy exists at the *interface* between the two phases (this is γ, the *interfacial tension*, or *surface tension*). Energy can also be applied from outside in the form of mixing (e.g., stirring by hand or beating with an electric mixer). Each time the oil or water droplets are subdivided by mixing, more droplets are created and the amount of interfacial area increases. The lower the interfacial energy is to start with, the larger the amount of new

Emulsion—A homogeneous dispersion of two dissimilar immiscible liquid phases. If oil is dispersed in water, it is an oil-in-water (O/W) emulsion. If water is dispersed in oil, it is a water-in-oil emulsion (W/O).

Creaming—In an emulsion, the collection of the lighter phase in the upper part of the mixture (e.g., oil droplets on top of water).

Emulsifier—A material that lowers the interfacial energy between two immiscible phases (e.g., oil and water), thus facilitating the dispersion of one phase into the other.

Surfactant—A chemical compound that concentrates at the interface between two dissimilar phases such as oil and water. The surface tension is lowered by the presence of a surfactant.

Interface—A surface that forms the common boundary between two bodies, spaces, or phases.

Interfacial tension—The forces that cause two dissimilar liquids such as oil and water to separate from each other. **Surface tension** is the force that causes two dissimilar phases such as air and water to separate. While sometimes used interchangeably, interfacial tension is between two liquids or solids while surface tension is between a liquid and a gas.

Amphiphilic—Describing a compound that possesses both lipophilic ("fat-loving") and hydrophilic ("water-loving") regions.

Lipophilic—Attracted to fat or nonpolar regions of molecules. This chemical property results from the occurrence of CH_2 groups and the absence of oxygen or nitrogen groups.

Hydrophilic—Attracted to water or polar regions of molecules. This chemical property results from the occurrence of oxygen or nitrogen groups and means that there is no attraction to fat or nonpolar groups.

interfacial area that can be created for a given amount of energy input. As the interfacial area increases, so does the interfacial energy. However, this is the energy that makes the droplets resist further subdivision. It makes the small droplets want to coalesce into larger droplets, leading eventually to separation of the oil and water phases. Added surfactant decreases the free energy at the oil-water interface, lowering interfacial tension and slowing the rate of coalescence. This chapter examines this process in more detail.

AMPHIPHILIC (POLAR AND NONPOLAR) NATURE

As mentioned above, an emulsifier is a type of surfactant. "Surfactant" is a coined word (from *surf*ace *act*ive ag*ent*) applied to molecules that migrate to interfaces between two physical phases and thus are more concentrated in the interfacial region than in the bulk phases. The key molecular characteristic of a surfactant is that it is *amphiphilic* in nature; the *lipophilic* (or hydrophobic) part of the molecule prefers to be in a lipid (nonpolar) environment and the *hydrophilic* part prefers to be in an aqueous (polar) environment (Box 3-1). If a surfactant is dissolved in either phase of an ordinary mixture of oil and water, some portion of the surfactant concentrates at the oil-water interface. At equilibrium, surface tension is lower than in the absence of the surfactant.

The lipophilic portion of food surfactants is usually a long-chain fatty acid obtained from a food-grade fat or oil (Fig. 3-1). The hydrophilic portion can be *nonionic*, *anionic* (negatively charged), or *amphoteric* (carrying both positive and negative charges). Cationic (positively charged) surfactants are usually bactericidal and somewhat toxic and are not used as food additives. Examples of the three types are monoglyceride (nonionic), stearoyl lactylate (anionic), and lecithin

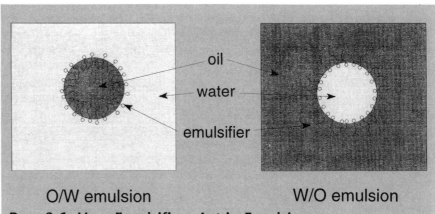

O/W emulsion   W/O emulsion

Box. 3-1. How Emulsifiers Act in Emulsions

In an oil-in-water (O/W) emulsion, drops of oil are suspended in water and molecules of emulsifier are gathered at the interface. The lipophilic part of the emulsifier (the long chain) is associated with the oil droplet, and the hydrophilic part is associated with the water. In a water-in-oil (W/O) emulsion, water droplets are dispersed in oil.

(amphoteric). The nonionic surfactants are relatively insensitive to pH and salt concentration in the aqueous phase, while the functionality of the ionic types may be markedly influenced by pH and ionic strength.

EMULSIFICATION

Promotion. The formation of an O/W emulsion is outlined in Figure 3-2. If 100 ml of pure vegetable oil plus 500 ml of water are mixed vigorously to obtain an emulsion in which the average diameter of the oil globules is 1 μm, then 600 m² (slightly more than 6,400 ft²) of oil-

Nonionic—Describing a compound with no positive or negative charge.

Anionic—Describing a negatively charged compound.

Amphoteric—Describing a compound such as a protein that has both positive and negative charges.

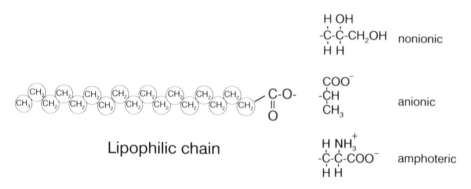

Fig. 3-1. Amphiphilic molecules. The fatty acid chain is the lipophilic (hydrophobic) part. Various hydrophilic end groups are shown.

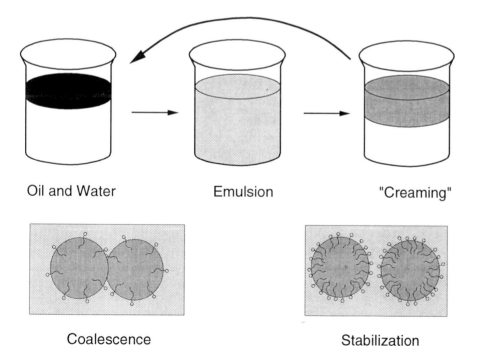

Fig. 3-2. Emulsification, creaming, coalescence, and stabilization of a mixture of oil and water.

Glycerol monostearate (GMS)—A monoglyceride made of one stearic acid molecule attached to glycerol.

water interface is generated. When a purified vegetable oil is used with water, γ (the interfacial tension) is about 30 erg/cm². To form the emulsion, 18 J of the energy input from the mixing is converted into interfacial energy. This increased interfacial energy is the driving force behind coalescence of the oil globules. The addition of 1% *glycerol monostearate* (GMS) to the oil phase lowers γ to about 3 erg/cm², making the excess interfacial energy only 1.8 J. In a series of emulsification experiments in which the amount of mixing energy was constant and γ was changed by adding emulsifier, the average oil droplet diameter was found to parallel γ; that is, as more emulsifier was added, γ decreased and so did average droplet size.

Stabilization. The phenomenon just described is the *promotion* of emulsion formation; this is not the same as the *stabilization* of emulsions. When two oil droplets make contact, they coalesce into one larger droplet (Fig. 3-2). If an O/W emulsion is allowed to stand, the oil droplets rise to the top. Often, this "creaming" is undesirable, as when oil separates from mayonnaise and comes to the top of the jar. The rate of creaming is inversely related to droplet diameter because large droplets rise faster than small droplets. An emulsifier that promotes the formation of a smaller droplet diameter yields an emulsion in which the rate of separation of oil and water is slower. In the same way, homogenization of milk reduces the average fat globule diameter from 10 μm to 1 μm, so the rate of creaming is greatly reduced.

When droplets coalesce into a larger droplet, total surface area and total excess interfacial energy are reduced. An emulsifier may be very efficient at reducing γ even at a low interfacial concentration, but it cannot prevent the oil droplets from touching and coalescing during creaming. A true emulsion stabilizer prevents the droplet contact that leads to coalescence.

Stabilization is of several types. One depends on electrostatic charges. For instance, if the surfactant is anionic, oil droplets carry a negative charge on their surfaces and thus repel each other by electrostatic effects. However, this sort of stabilization is sensitive to ionic strength: a high salt concentration suppresses the electrostatic repulsion, promotes contact and coalescence, and leads to rapid breakdown of the emulsion. A second kind of stabilization is accomplished by surfactants in which the hydrophilic portion is quite large, for example, ethoxylated monoglyceride (EMG) or the polyoxyethylene chain of polysorbate. In this case, the chain is anchored at the surface of the oil droplet by the lipophilic tail, but it is strongly hydrated and generates a layer of "bound" water around the droplet, preventing contact and coalescence. This functionality is relatively insensitive to salt concentration. Yet a third kind of stabilization is due to simple steric hindrance of contact. Certain emulsifiers (called α-tending emulsifiers; see section on film formers later in this chapter) such as propylene glycol monostearate (PGMS) form an actual solid layer at the oil-water interface. This film physically prevents the oil contained in droplets from coalescing even though their surfaces may be touching.

It should be noted that emulsion stabilization is not directly related to the ability to lower interfacial tension. GMS (1%) in the oil phase lowers γ to 3 erg/cm² but has little effect on coalescence rate, while 12% PGMS yields a γ of 9 erg/cm² but gives an emulsion with excellent long-term stability.

FOAMING

Whereas an emulsion is a mixture of two immiscible liquids (e.g., water and oil), a *foam* is a mixture of liquid and gas (e.g., water and air). It involves the same energy considerations as an emulsion: lowering the interfacial (surface) tension favors foam formation. Small surfactant molecules dissolved in the aqueous phase promote foaming, and the stability of the foam is dependent on the stability of the film of water between air bubbles. The lipophilic portion of the surfactant enters the gas phase (rather than an oil phase), but in all other respects the situation is analogous to that in emulsions.

The foaming agent in foods is often a protein. Proteins are amphiphilic molecules; many amino acids have hydrophobic (lipophilic) side chains, and others have side chains that are ionic, hence hydrophilic. Normally, proteins such as egg albumin in solution are folded in such a way that the hydrophobic side chains are buried in a nonpolar environment in the interior of the molecule, while the hydrophilic side chains are on the surface of the molecule and interact with the polar aqueous environment. When air bubbles are introduced into the solution by the action of the mixer, the protein unfolds at the air-water interface. The hydrophobic side chains enter the air phase, and the hydrophilic chains remain in the water phase. The portions of the proteins located in the aqueous phase hold water, prevent it from draining, and prevent the air bubbles from coalescing and destabilizing the foam.

Whipping aids increase the ability of the protein to unfold at the air-water interface. This effect may be due to denaturation caused by a pH change (e.g., from cream of tartar) or to the presence of a surfactant (e.g., sodium lauryl sulfate). In either case, the energetics of protein unfolding become more favorable and the ease of foam formation increases. On the other hand, the presence of lipids retards foaming because the oil molecules migrate to the air-water interface before the protein molecules, inhibiting the unfolding of the protein and thus the formation of foam.

WETTING

In some surfactants, the lipophilic and hydrophilic tendencies are evenly balanced, so these materials do not favor the subdivision of either the oil or water phase in a mixture. The *hydrophilic-lipophilic balance (HLB)* of such emulsifiers is usually around 8–10 (see section on HLB below). They promote the spreading of a liquid phase onto a solid surface, for instance, in the mixing of water and cocoa powder. Generally the surface is somewhat hydrophobic (e.g., cocoa solids), and the emulsifier increases the rate at which water displaces air from the sur-

Foam—A gaseous phase, such as air, dispersed and held in a liquid phase, such as water.

Hydrophilic-lipophilic balance (HLB)—A system of classifying surfactants or emulsifiers by how much "water loving" and how much "fat loving" character the molecules have. On a scale of 0–20, a high number means more hydrophilic.

face of the particles, enhancing the dispersion of the solid phase into the water. This functionality is useful in formulating dry drink mixes, coffee creamers, and similar food products.

INTERACTIONS WITH OTHER COMPONENTS

Although starch and proteins are usually thought of as being hydrophilic, readily wetted by water and becoming water soluble as the molecular size is reduced, in fact they are amphiphilic. As such, they interact with surfactants.

Starch *gelatinization*. Native starch granules are ordered structures possessing a significant degree of crystallinity. When starch is heated in water, this crystalline property begins to disappear at a temperature characteristic of the source of the starch and is almost completely lost at a point about 6–8 degrees C higher. This loss of crystallinity is part of the process known as gelatinization.

Surfactants modify the gelatinization behavior of starch. Figure 3-3 shows the changes in amylograph curves during gelatinization of wheat starch that were caused by the inclusion of 0.5% of various emulsifiers. Of the three emulsifiers shown, diacetyl tartaric acid esters of monoglyceride (DATEM) is the least interactive, raising the swelling temperature by about 5 degrees C but not changing the viscosity of the gelatinized starch. GMS causes the most dramatic changes, raising swelling temperature by about 18 degrees C and also increasing the paste viscosity. Sodium stearoyl lactylate (SSL) is less effective in delaying swelling temperature, but it increases paste viscosity to about the same extent as does GMS.

Protein *denaturation*. As discussed above, some of the amino acid side chains in proteins are hydrophobic, generally buried in the interior of the native protein molecule but exposed if the protein is unfolded. Sometimes these hydrophobic regions are partially exposed even in the native, folded, protein. The lipophilic parts of surfactants interact with these hydrophobic regions, contributing to unfolding (denaturation) of the protein and further binding of the surfactant.

An excess amount of surfactant can solubilize proteins, presumably by additional adsorption to the surface of the protein chain. This generates a large net charge at the surface of the protein molecule. This is the basis for estimation of protein molecular weight by polyacrylamide gel electrophoresis in the presence of sodium dodecyl sulfate (SDS-PAGE). Such effects are usually found only at surfactant concentrations well in excess of those normally found in foods, and they have more use in laboratory investigations than in real applications.

Gelatinization—The process in which starch molecules swell and lose crystallinity in the presence of water and heat.

Denaturation—Any process in which protein molecules irreversibly change their native shape as a result of forces such as heat, agitation, or solvents.

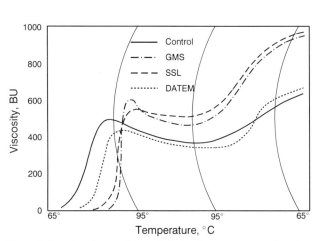

Fig. 3-3. Amylograph gelatinization curves showing emulsifiers affecting starch gelatinization in different fashions. Some raise the gelatinization temperature, while others affect the viscosity of the starch gel. The emulsifiers shown were used at a concentration of 0.5% relative to the starch. GMS = glycerol monostearate, SSL = sodium stearoyl lactylate, DATEM = diacetyl tartaric ester of monoglyceride.

Functional Properties

EMULSIFICATION—HLB

The type of emulsion formed is determined by the solubility characteristics of the emulsifier. As a general rule, the continuous phase is the one in which the emulsifier is soluble. Thus, a soap (e.g., sodium stearate) promotes an O/W emulsion, while monoglyceride promotes a water-in-oil (W/O) emulsion. The concept of hydrophilic-lipophilic balance, or HLB, as a number for characterizing surfactants is an extension of this general rule. The scale as originally proposed ranged from 0 to 20, the low end signifying an emulsifier that is much more soluble in oil than in water, and the high end signifying the reverse. Table 3-1 lists HLB values for some food surfactants. HLB is recommended as a guide for selecting emulsifier systems, especially for foods such as salad dressings, whipped toppings, and similar oil/water mixtures that must retain their emulsified character during a long shelf life, up to one year.

A basic part of the concept is that the HLB of a blend of two emulsifiers is equal to their algebraic sum, i.e., the weight fraction of emulsifier A times its HLB value plus the weight fraction of emulsifier B times its HLB value. In a test system comprising equal volumes of water and Nujol (a food grade mineral oil), using blends of various Span (sorbitan fatty acid esters) surfactants and Tween (polyoxyethylene$_{20}$ sorbitan fatty acid esters) surfactants to cover the HLB ranges from 2.0 to 6.5 and 8.5 to 16.5, O/W emulsions were most stable at an HLB of about 12 and W/O emulsions at an HLB of about 3.5. In a practical system that includes other ingredients such as sugar, salt, protein, and other typical food components, the optimal HLB might be somewhat different, and a series of tests is necessary to determine the exact blend of surfactants that gives the best results. As a rule of thumb, W/O emulsions are stabilized by HLB values in the 3–6 range, O/W emulsions are stabilized by HLB values in the 11–15 range, and intermediate HLB values give good wetting properties but not good emulsion stability.

ANTISTALING— COMPLEXING WITH STARCH

Starch molecules are polymers composed of glucose units (α-D-glucopyranosidyl residues) joined primarily by 1,4 acetal linkages (1,6 linkages occur at the branch points in amylopectin). The starch glucopyranoside (glucose) is a six-membered ring in the "chair" configuration (Fig. 3-4A). The hydroxyl groups project outward to the side of the plane of the ring, while the hydrogen atoms project either above or below this plane. The perimeter of the ring is hydrophilic, while the two faces are hydrophobic.

TABLE 3-1. Hydrophilic/Lipophilic Balance (HLB) Values for Some Food Surfactants

Surfactant	HLB Value
Sodium stearoyl lactylate (SSL)	21.0
Polysorbate 80 (PE$_{20}$ sorbitan monooleate)	15.4
Polysorbate 60 (PE$_{20}$ sorbitan monostearate)	14.4
Sucrose monostearate	12.0
Polysorbate 65 (PE$_{20}$ sorbitan tristearate)	10.5
Diacetyl tartaric ester of monoglyceride (DATEM)	9.2
Sucrose distearate	8.9
Triglycerol monostearate	7.2
Sorbitan monostearate	5.9
Succinylated monoglyceride (SMG)	5.3
Glycerol monostearate (GMS)	3.7
Propylene glycol monoester (PGME)	1.8

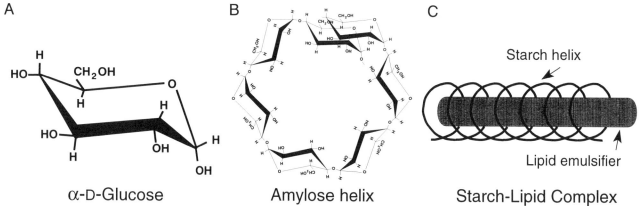

Fig. 3-4. Configuration of: A, α-D-glucopyranoside; B, the helix formed by glucopyranoside linked by α-1,4 bonds; C, the complex between a fatty acid chain and the helix.

Helix—A linearly coiled structure with a regular pattern of turns.

In solution, the starch chain coils to form a *helix*, with about six residues per turn (Fig. 3-4B). The result is a hollow cylinder with a hydrophilic outer surface and a hydrophobic inner surface. The inner space is about 4.5 Å in diameter, and straight-chain alkyl molecules such as stearic acid can fit into it (Fig. 3-4C).

In this way, the fatty acid part of emulsifiers such as GMS can form a complex with gelatinized starch, retarding starch crystallization in bread crumb and slowing the process of staling. The amount of monoglyceride that complexes with *amylose* (the linear component of starch) and with *amylopectin* (the branched starch component) is shown in Figure 3-5. (Other studies report different monoglyceride-starch ratios, but the pattern is the same.) Unsaturated fatty acids have a bend produced by the double bond in the fatty acid chain, which limits their ability to form the complex. The overall picture is that the straight-chain hydrophobic portion of emulsifiers complexes with helical sections of amylose and amylopectin, as shown in Figure 3-4C.

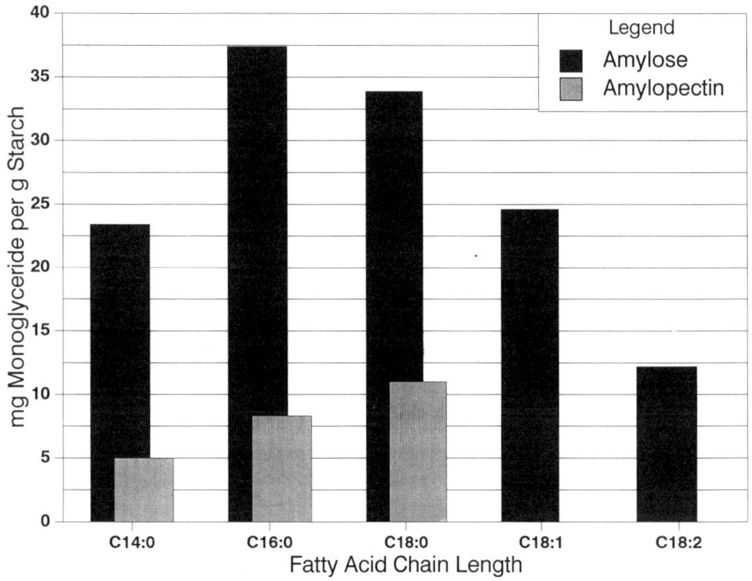

PROTEIN AGGREGATION AND SOLUTION

As mentioned earlier, sometimes hydrophobic regions of the amino acid side chains in proteins are partially exposed, even in the folded protein. These are often referred to as "hydrophobic patches" on the protein surface.

Fig. 3-5. Amount of monoglyceride complexed by amylose and amylopectin. Various monoglycerides react to different extents, and saturated fatty acids form a better complex than unsaturated fatty acids.

The lipophilic parts of surfactants may interact with these hydrophobic regions.

Wheat gluten protein consists of about 40% hydrophobic amino acids. Consequently, it interacts with lipid-type materials such as the surfactant SSL. Figure 3-6 depicts the effect of this interaction on gluten characteristics. The addition of acid to a flour-water dough solubilizes some of the protein. Dough pH (about 6) is roughly the *isoelectric point* of the gluten proteins (Fig. 3-6A). As the pH is lowered by the acid, many of the organic acid groups are *protonated*, i.e., become neutral, and the protein molecule takes on a net positive charge (Fig. 3-6B). At pH 6, the hydrophobic patches can interact, and the protein molecules aggregate via hydrophobic interaction, but at pH 3, the net positive charges cause the molecules to repel each other, creating a dispersion. (This situation has many similarities to the stabilization of emulsified oil droplets by an ionic surfactant, mentioned earlier, in which the surface charge prevents droplet contact and coalescence.) If an emulsifier is then added, the protein molecules can aggregate even at pH 3 (Fig. 3-6C). Most dough strengtheners are anionic surfactants. When the lipophilic tail of the surfactant binds to the hydrophobic patches on the protein molecule's surface, it incorporates this negative charge into the complex, moving the net charge closer to zero and promoting protein (gluten) aggregation in the dough (Fig. 3-6C).

Amylose—The type of starch molecule that occurs as a linear coil without branching.

Amylopectin—The type of starch molecule that has branches.

Isoelectric point—The pH level at which the number of positive charges is equal to the number of negative charges (e.g., on a protein).

Protonated—Describing a chemical group that can reversibly add a hydrogen ion (proton) to its structure. If the hydrogen is present, the chemical group is in the protonated form. COOH is the protonated form of COO⁻.

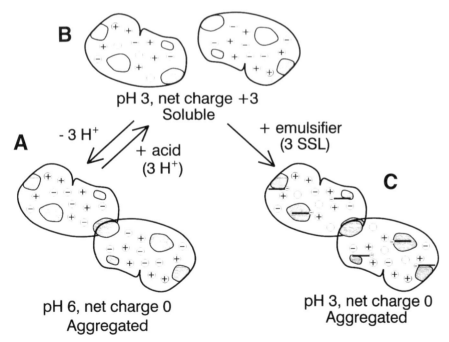

Fig. 3-6. Model for the aggregation of gluten molecules, showing changes in solubility brought about by protons (left) or by sodium stearoyl lactylate (SSL, right). A, dough, with aggregated protein molecules at pH 6; B, unaggregated protein molecule at lowered pH level; C, aggregated molecules at lowered pH level.

Emulsifier Types and Applications

EMULSIFIERS

Emulsified shortening—A shortening that is manufactured with added emulsifiers (surfactants).

Staling—The firming of baked products that occurs during storage. The loss of soft texture and flavor is related to changes in the starch after baking.

Distilled monoglycerides—A preparation of monoglycerides that is prepared by distillation to separate the monoglycerides from other components, primarily diglycerides.

Hydrated monoglycerides—A monoglyceride preparation that contains monoglycerides, water, and other surfactants to keep the mixture homogeneous.

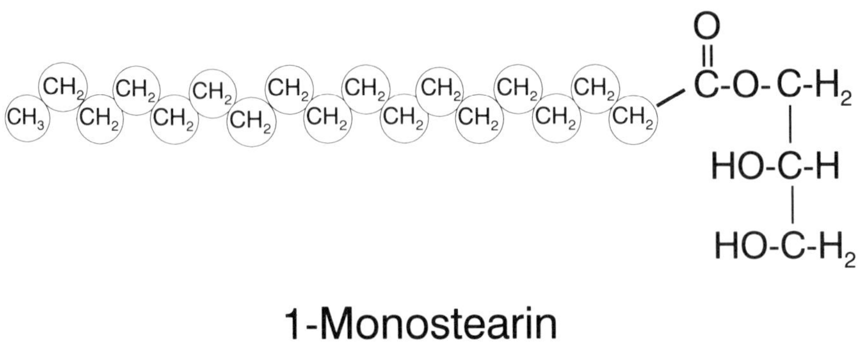

Monoglyceride. Because of its ability to retard staling, roughly 40 million pounds of monoglyceride is used annually in yeast-raised goods in the United States. At least an equal amount finds its way into cake icings (where it stabilizes the incorporation of air into the shortening) or is used in the manufacture of margarine (for promoting a W/O emulsion during manufacture). Overall, this group of surfactants is the single most important one for food uses.

Superglycerinated shortening, introduced in the 1930s, was made by adding glycerin to ordinary shortening and heating with a catalyst. The resulting *emulsified shortening* contained about 3% monoglyceride and was widely used for making high-ratio cakes (i.e., cakes containing more sugar than flour). The effectiveness of monoglyceride in retarding *staling* (crumb firming) in bread became known at about the same time. A more concentrated plastic monoglyceride is made by increasing the ratio of glycerin to fat to achieve a final concentration of 50–60% monoglyceride, with most of the remainder being diglyceride.

In the 1940s, molecular distillation methods were developed to produce *distilled monoglyceride* containing a minimum of 90% monoglyceride. *Hydrated monoglyceride*, made from distilled monoglyceride, contains roughly 25% monoglyceride, 72% water, and 3% SSL (which stabilizes the hydrate). A powdered distilled monoglyceride contains a balance between saturated and unsaturated fatty acids (the iodine value is around 20), so the resulting powder is hydrated fairly rapidly during dough mixing. The uses of these various forms of monoglyceride in baking are discussed more fully in Chapter 5.

The monoglyceride structure shown in Figure 3-7 is for 1-monostearin, also called α-monostearin. If the fatty acid is esterified at the middle hydroxyl, the compound is 2-monostearin, or β-monostearin. Both are isomers of GMS and both are equally effective at retarding bread staling and in other food applications. The routine analytical method for monoglyceride detects only the 1-isomer. In a practical sense, however, when various products are being compared for functionality and cost effectiveness, the α-monoglyceride content is a

1-Monostearin

Fig. 3-7. A typical monoglyceride, 1-monostearin.

useful number since, for all products, this equals about 92% of the total monoglyceride present.

The fatty acid composition of a monoglyceride reflects the makeup of the triglyceride fat from which it was made. A commercial preparation of GMS may contain as little as 65% stearate if it is made from fully hydrogenated lard or as much as 87% stearate if it is made from fully hydrogenated soybean oil. The other major saturated fatty acid in GMS is palmitic, and, because hydrogenation is never complete (the iodine value of GMS is about 5), a few percent of unsaturated (oleic and/or elaidic) fatty acid is also present. Iodine values for powdered distilled monoglycerides are in the range of 19–36, and for plastic monoglycerides a typical range is 65–75.

Polysorbates, **sorbitan**. Heating sorbitol (a sugar) with stearic acid in the presence of a catalyst cyclizes the sorbitol and forms an ester, to produce sorbitan monostearate (Fig. 3-8A). Other sorbitan esters of importance are the monooleate and the tristearate. Any of the three esters may be reacted with ethylene oxide to give polyoxyethylene derivatives (Fig. 3-8B and C), which are much more hydrophilic than the sorbitan esters. The monostearate derivative is known as Polysorbate 60; the tristearate is Polysorbate 65; and the monooleate is Polysorbate 80. The average number of oxyethylene monomers is 20 ($n = 20$). In the case of

Polysorbates—A group of emulsifiers that each contain sorbitans, various types and amounts of fatty acids, and polyoxyethylene chains.

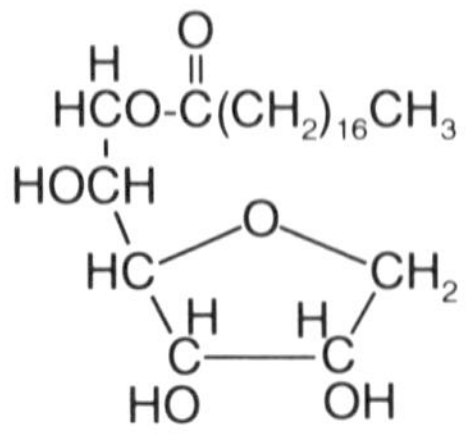

Sorbitan monostearate

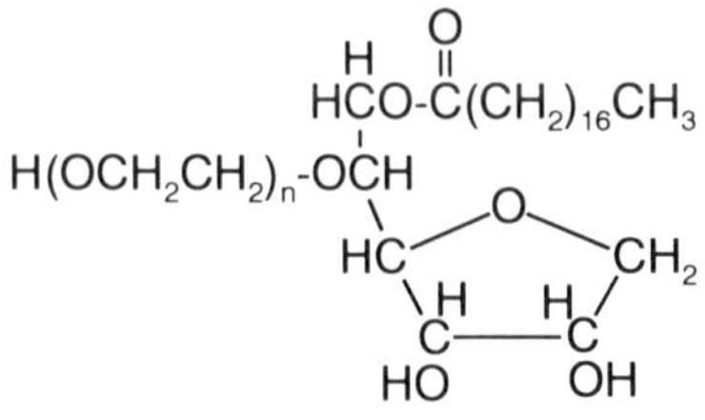

Polyoxyethylene (20) sorbitan monostearate (Polysorbate 60)

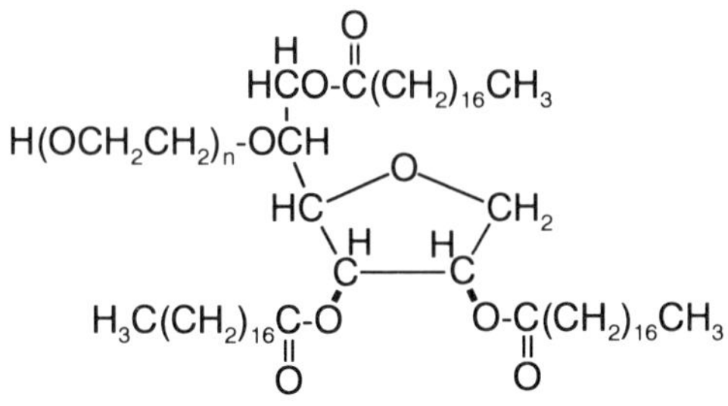

Polyoxyethylene (20) sorbitan tristearate (Polysorbate 65)

Fig. 3-8. Structure of sorbitan monostearate and two polysorbates.

Polyglycerols—A group of emulsifiers that contain polymerized glycerol and various types and amounts of fatty acids.

Esterification—The formation of ester bonds by joining hydroxyl groups on glycerol or sugars with carboxylic acids on fatty acids.

Sucrose esters—Emulsifiers manufactured by adding fatty acids to a sucrose molecule.

the monoesters, chains may be located on more than one hydroxyl group of the sorbitan ring. These emulsifiers have a broad range of uses in foods.

Polyglycerol esters. Polyglycerol esters (Fig. 3-9A) have a variety of applications as emulsifiers in the food industry. The polyglycerol portion is synthesized by heating glycerol in the presence of an alkaline catalyst; ether linkages are formed between the primary hydroxyls of glycerol. In Figure 3-9, n (an average value for all species present) may take any value, but for food emulsifiers, the most common are $n = 1$ (triglycerol), $n = 4$ (hexaglycerol), $n = 6$ (octaglycerol), and $n = 8$ (decaglycerol). The polyglycerol backbone is then *esterified* either by direct reaction with a fatty acid or by interesterification with a triglyceride fat.

The HLB values of these esters depend upon the length of the polyglycerol chain (the number of hydrophilic hydroxyl groups present) and the degree of esterification. As examples, decaglycerol monostearate has an HLB value of 14.5, while triglycerol tristearate has an HLB value of 3.6. Intermediate HLB values may be obtained by appropriate blending, as described earlier. The wide range of possible compositions and HLB values makes these materials versatile emulsifiers for food applications.

Sucrose esters. Sucrose has eight free hydroxyl groups that are potential sites for esterification to fatty acids. Derivatives containing one to three fatty acid esters (Fig. 3-9B) are emulsifiers and are approved for food use. They are manufactured by the following steps: 1) an emulsion is made of fatty acid methyl ester in a concentrated aqueous sucrose solution; 2) the water is removed under vacuum at elevated temperature; 3)

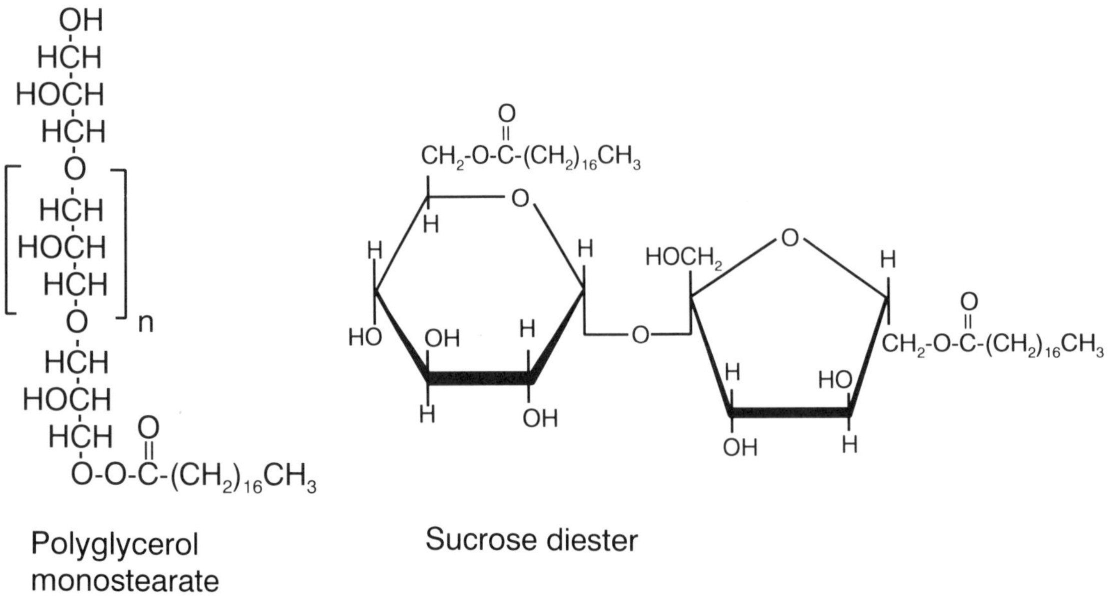

Polyglycerol monostearate

Sucrose diester

Fig. 3-9. Polyglycerol esters and sucrose esters.

TABLE 3-2. Sucrose Ester Surfactants

Manufacturer's Designation	Percent Monoester	Percent Diester	Percent Triester	Percent Tetraester	HLB[a] Value
F-160	71	24	5	0	15
F-140	61	30	8	1	13
F-110	50	36	12	2	11
F-90	46	39	13	2	9.5
F-70	42	42	14	2	8
F-50	33	49	16	2	6

[a] Hydrophilic-lipophilic balance.

alkaline catalyst is added, and the temperature of the dispersion is raised slowly to 150°C under vacuum, distilling off the methanol formed upon transesterification; and 4) the reaction mixture is cooled and purified. The degree of esterification is controlled by the reaction conditions, especially the sucrose-methyl ester ratio, and the final product is a mixture of esters (Table 3-2). The HLB value of a particular product is lower (more lipophilic) as the degree of esterification increases. As with the sorbates and polyglycerol esters, the wide range of possible HLB values allows use of sucrose esters in many different food applications.

WETTING AGENTS

Lecithin. Commercial lecithin is usually a by-product (often termed a "gum") from the refining of crude soybean oil. Figure 3-10 shows the main surface-active components of soy lecithin, which are present in approximately equal amounts. Phosphatidylethanolamine (PE) and phosphatidylcholine (PC) are amphoteric surfactants, and phosphatidylinositol (PI) is anionic. PC has a high HLB value, PE an intermediate HLB value, and PI a low HLB value. The HLB of the natural blend is in the range of 8–10, forming either O/W or W/O emulsions, although neither type is very stable. Intermediate-HLB emulsifiers are excellent wetting agents, and this is a major application for lecithin.

The emulsifying properties of lecithin can be improved by ethanol fractionation. PC is soluble in ethanol; PI is rather insoluble; and PE is partially soluble. Adding lecithin to ethanol produces a soluble and an insoluble fraction. The phosphatide compositions of the two are: 1) ethanol-soluble fraction: 60% PC, 30% PE, and 2% PI and 2) ethanol-insoluble fraction: 4% PC, 29% PE, and 55% PI. The soluble fraction is a good O/W emulsifier, while the insoluble portion stabilizes W/O emulsions.

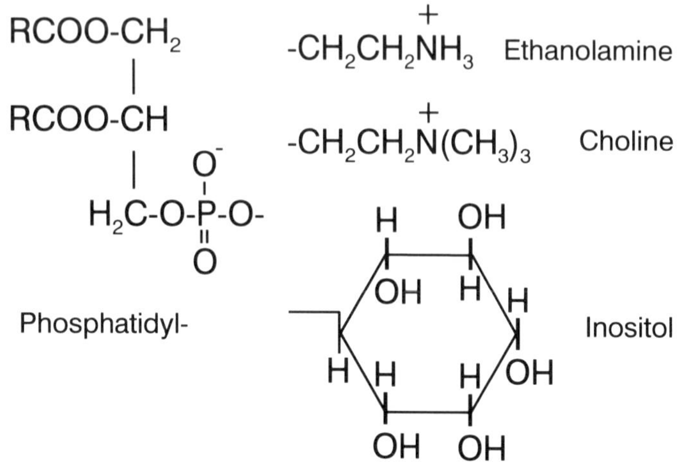

Fig. 3-10. The main components of soy lecithin. Each compound indicated is present at about 30% of the total weight of the phospholipid. Other minor components include lysolecithins (having only one fatty acid chain) and phosphatidylserine.

Sodium lauryl sulfate (SLS), triethyl citrate, and dioctyl sodium sulfosuccinate. SLS and triethyl citrate are added to dry egg whites to improve their dispersion in water and their whippability in making meringues and angel food cakes. SLS is approved for use as a wetting agent for dry gelatin desserts and beverage bases that are acidulated with fumaric acid. Dioctyl sodium sulfosuccinate is used as a wetting agent for gelatin desserts, beverage bases, dry food gums, and cocoa intended for further manufacturing use.

DOUGH STRENGTHENERS

Dough strengthener—An ingredient such as a surfactant that is added to a dough in low levels to improve the dough properties.

A *dough strengthener* is added to bread formulas to improve dough machinability, bread quality, and loaf volume. A test of the dough-strengthening capabilities of a surfactant is to subject the proofed loaf to mechanical abuse before putting it into the oven. For example, the pan containing the proofed dough is dropped three times (from about 4 in.) onto the countertop. Then the doughs are baked, depanned, and cooled, and the volume is measured. This test is designed to simulate the rough handling that proofed doughs often receive in a commercial bakery, such as sudden starts and stops on the conveyor line or being struck by a loading or unloading bar. In one test, using surfactant at a level of 0.25% (flour basis), calcium stearoyl lactylate (CSL) gave the best volume improvement (approximately 260 cm^3 in a 1-lb loaf). Ethoxylated monoglyceride (EMG) was the second most effective; Polysorbate 60 and SSL were about equivalent (about 210 cm^3); and succiny-

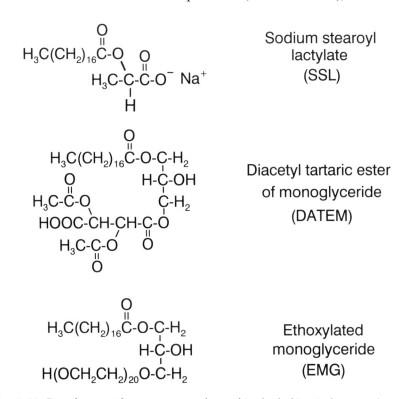

Fig. 3-11. Dough strengtheners commonly used in the baking industry today.

lated monoglyceride (SMG) gave the least improvement (about 125 cm³), all relative to the control.

Stearoyl lactylate (as the sodium and/or calcium salt) is the most prevalent dough strengthener used in U.S. bakeries today (Fig. 3-11). In Europe, DATEM is the main dough strengthener used and apparently is equally satisfactory. The difference appears to be the result of history (SSL was developed by a U.S. firm, DATEM by a European one) and marketing. At the present time, DATEM is being promoted in the United States and SSL in Europe. SSL (but not CSL) gives somewhat more antistaling effect than does DATEM, but, other than that, no information about functionality has been published that gives the baker a compelling reason to choose one surfactant over the other. Polysorbate 60 and EMG have shown particular effectiveness in high-fiber (and high-moisture) bread doughs.

FILM FORMERS

Unlike most fats, in which the α crystal phase readily transforms to β′, certain emulsifiers solidify in a stable α-crystalline (waxy) form. The main examples in commercial use (Fig. 3-12) are acetyl-monoglycerides (AcMG), lactyl-monoglycerides (LacMG), and propylene glycol monoesters (PGME). At concentrations above a certain level, these *α-tending emulsifiers* form a solid film at the oil-water interface. This phenomenon, and its usefulness in cake production, is discussed in Chapter 5. The addition of a second surfactant may enhance film formation; a mixture of PGME and stearic acid (80:20) is a stronger film-former than pure PGME at the same weight concentration.

α-Tending emulsifiers—Oil-soluble emulsifiers that form a solid film at the oil-water interface.

Acetyl-monoglyceride
(AcMG)

Lactyl-monoglyceride
(LacMG)

Propylene glycol
monostearate
(PGMS)

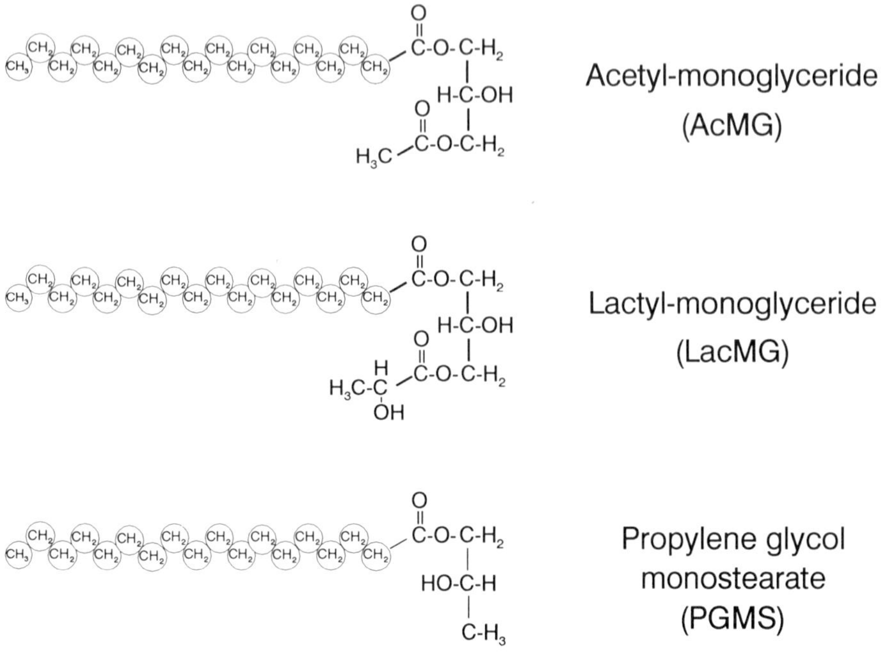

Fig. 3-12. Structure of several α-tending emulsifiers that form solid films at the oil-water interface under the proper conditions of temperature and concentration.

TABLE 3-3. Regulatory Status of Emulsifiers

Emulsifier	US 21 CFR	Canadian[a]	EU No.
Mono- and diglycerides (GRAS)[b]	182.4505	M.4, M.5	E 471
Succinyl monoglyceride	172.830		
Lactylated monoglyceride	172.852	L.1	E 472
Acetylated monoglyceride	172.828	A.2	E 472
Monoglyceride citrate	172.832		E 472
Monoglyceride phosphate (GRAS)	182.4521	A.94, C.7	
Stearyl monoglyceride citrate	172.755		E 472
Diacetyl-tartrate ester of monoglyceride (GRAS)	182.4101	A.3	E 472
Polyoxyethylene monoglyceride	172.834		
Polyoxyethylene (8) stearate		P.5	
Propylene glycol monoester	172.854	P.14	E 477
Lactylated propylene glycol monoester	172.850		
Sorbitan monostearate	172.842	S.18	E 491
Sorbitan tristearate		S.18B	
Polysorbate 60	172.836	P.3	E 435
Polysorbate 65	172.838	P.4	E 436
Polysorbate 80	172.840	P.2	E 433
Calcium stearoyl lactylate	172.844		E 482
Sodium stearoyl lactylate	172.846	S.15A	E 481
Stearoyl lactylic acid	172.848	L.1A	
Stearyl tartrate			E 483
Stearyl monoglyceridyl citrate	172.755	S.19	
Sodium stearyl fumarate	172.826		
Sodium lauryl sulfate	172.822		
Dioctyl sodium sulfosuccinate	172.810		
Polyglycerol esters	172.854	P.1A	E 475
Sucrose esters	172.859	S.20	E 473
Sucrose glycerides			E 474
Lecithin (GRAS)	184.1400	L.2	E 322
Hydroxylated lecithin	172.814	H.1	E 322
Triethyl citrate (GRAS)	182.1911		

[a] Canadian Food and Drug Regulations, Table IV, Div. 16.
[b] Generally recognized as safe.

Regulatory Considerations

Food additive—A food ingredient category in which certain restrictions on use apply (a regulatory term).

Emulsifiers are regulated as *food additives* in most countries. The relevant U.S., Canadian, and European Union regulatory references are listed in Table 3-3 for most food surfactants. In many cases, the maximum permissible amount of emulsifier is specified; in other instances, the manufacturer is expected to use no more than the amount necessary to provide the desired technological effect. If there is any question about the use of an emulsifier in a particular application, the appropriate document must be consulted.

Refining and Production

Processing and Refining

Fats and oils as they exist in nature must be processed before they are suitable for use as edible fats and oils. In some cases, this is minimal, for example, churning cream to obtain butter. In other instances, the route from natural product to edible oil is much lengthier, as in making salad oil from soybeans or cottonseed. *Processing* isolates the fat or oil from its natural matrix, while *refining* removes impurities and gives a fully edible product.

VEGETABLE OILS

Crushing, extracting. All plant cells and seeds contain a certain amount of fat, in addition to carbohydrates and protein, which is necessary to sustain their growth. Generally speaking, the methods of recovering oil follow very simple principles. The oil is either squeezed from the seed using a press (hydraulic, screw, or a combination), or it is extracted using a solvent. Frequently, the two methods are combined, depending on the nature of the seed and the cost of the operation. The prime objective in using any of these methods is to obtain the maximum amount of an unaltered oil that is as free of impurities as possible.

Most oil seeds, except for rapeseed (canola), are dehulled before extraction. This makes it easier to extract oil from the meats and increases yield by decreasing oil absorption by the hulls. The meats are flaked, causing sufficient cell breakage to release the oil for ease of extraction (Figs. 4-1 and 4-2). Mechanical extraction is simpler (and safer) but less efficient than solvent extraction. It leaves a product (*press meal*) that contains 3–6% oil, whereas solvent-extracted meal contains 0.5–1.5% oil. In solvent extraction, the flakes (or press meal) are soaked in the solvent, normally hexane (a light petroleum fraction). The recovered solvent-oil mixture is called "*miscella,*" and the extracted flakes are referred to as "*spent flakes.*" These often have more economic value than the oil, being used as a high-protein animal feed or as a source of isolated protein for human food.

Refining. Oilseeds are natural, biologically active materials that contain many color and flavor precursors as well as degradation and breakdown products. Some of the extraneous materials in crude oil are phosphatides (gums), free fatty acids, color compounds (carotenoids and chlorophyll), tocopherols (vitamin E), waxes, moisture, meal fines, and dirt. The oil must be refined to remove these impurities.

Processing—Removing or otherwise extracting a fat or oil from its natural matrix.

Refining—Removing impurities from an extracted fat or oil.

Press meal—The material remaining when oil is extracted by mechanical means from oil seeds.

Miscella—The mixture of solvent and oil that results from the solvent extraction of oil from oil seeds.

Spent flakes—The material remaining when the oil is removed from oil seeds.

Degumming. Dry gums are soluble in crude oil but become oil-insoluble upon hydration. The degumming process is enhanced, and the rehydration rate is accelerated, by the addition of phosphoric acid (and sometimes citric acid) to the water. After mixing and contact time sufficient to hydrate the gums, these are separated by centrifugation. Good control of the amount of water and acid, the temperature, and the length of contact time is required to achieve the maximum extent of degumming, while minimizing oil losses and maximizing product quality.

Neutralization. Crude oils contain free fatty acids from natural hydrolysis of the triglycerides. These are removed by the addition of aqueous sodium hydroxide solution, which converts the fatty acids to water-soluble soaps. The soap solution is removed, either by simply draining it off or by centrifugation. After the solution is removed, the oil is washed once or twice with water to ensure the removal of all traces of soap and is dried.

Dewaxing. Sunflower and corn oils contain relatively large amounts of waxes that must be removed so that the oil does not cloud when refrigerated. The oil is cooled to 5°C, mixed with water containing a surfactant, and allowed to stand for several hours. If neutralization is done at this temperature, the soaps wet the wax crystals and bring them into suspension in the aqueous phase. Alternatively, sodium lauryl sulfate is added to the water to serve the same purpose. The aqueous wax suspension is removed by centrifugation.

Bleaching. Colored substances and oxidation products are removed by bleaching. Bleaching earths are natural clays (bentonites) treated with

Bleaching—Removing colored substances from an oil by absorbing them onto a solid material such as clay.

Fig. 4-1. Soybeans at several stages of processing. (Courtesy Buhler Inc., Minneapolis, MN)

acid to improve their adsorptivity and filterability. Proper bleaching with acid-activated earth is one of the most critical steps in oil processing. When done correctly, the oil emerging from the bleaching press is nearly colorless, with a peroxide value of zero.

Filtration. Complete removal of bleaching earth from the oil by effective filtration is very important because residual clay acts as a very strong prooxidant and can foul the equipment downstream. Some refineries still use the old plate and frame filter presses, which are very effective pieces of equipment. However, more modern filtering equipment is on the market. The recovered bleached oil should be protected against thermal and oxidative abuses, as the oil, at this point in the refining process, is in its least stable state. The headspace in storage tanks should be filled with nitrogen gas to exclude oxygen and prevent autoxidation.

Deodorization. If the oil is to be used as a salad oil, numerous compounds causing undesirable odors and flavors must be removed. These compounds are relatively volatile and can be removed by bubbling live steam through the hot oil (200–275°C) under high vacuum (3–10 mm Hg). Deodorization does not have any significant effect upon the fatty acid composition of most fats and oils. The oil at this step is referred to as *refined, bleached, deodorized (RBD) oil.* Deodorization removes free

Deodorization—Removal of odors from an oil by injecting steam. The undesirable odors are carried away because of their volatility.

RBD oil—Refined, bleached, and deodorized oil. These three treatments are frequently applied in series to convert extracted oils into more desirable products.

Fig. 4-2. Flaking mills. (Courtesy Buhler Inc., Minneapolis, MN)

fatty acids, but the live steam also causes some hydrolysis. Under optimum conditions, at equilibrium the oil contains 0.02–0.04% free fatty acids and 0.3–0.5% monoglycerides.

If the oil is to be hydrogenated to make basestocks for shortening or margarine manufacture, the deodorization step is often skipped at this point and the oil is stored under nitrogen. The blended basestock oils are deodorized just before being chilled to make the plastic fat (see section on blending of basestocks below).

ANIMAL FATS

Adipose tissue. Nearly all animal fats (lard, tallows) are recovered by rendering, which is the extraction of fat or oil by heat. The two general methods are dry and wet rendering.

In *dry rendering*, the tissues are heated and dehydrated until brittle, and the fat cells break and release the fat. The frying of bacon, to cite a familiar example, is essentially a dry rendering process. This method is typically used for inedible products in which flavor and color are secondary and the production of larger quantities of high-quality residue is the prime objective. Final recovery is completed by pressing the residue, either in hydraulic or continuous-screw presses.

In *wet rendering*, the fat cell walls are destroyed by steam under pressure, and the released fat is skimmed from the top of the water or separated by centrifugation. This method is used to obtain edible fat products in which color, flavor, and keeping qualities are of prime importance.

Lard is usually not deodorized. The best grades are produced by wet rendering and have a mild, unobjectionable flavor. Tallow, on the other hand, is generally dry rendered and has a strong meaty flavor, which is removed by steam deodorization.

Milk fat. The process of converting fat globules found in the milk of domesticated animals to an edible fat food is of great antiquity. Butter is a water-in-oil (W/O) emulsion containing about 20% water and therefore is susceptible to mold growth. In many tropical regions, the butter is melted and the water removed to give a food oil called *ghee*, which is more stable in storage than butter. Although butter consumption in the United States has decreased in the past 50 years, worldwide it is still the number two edible fat, exceeded only by soybean oil in quantities consumed.

The fat globules in milk are stabilized by a surface layer of lipoprotein. Cream (at 40% fat content) is agitated in a churn, which ruptures lipoprotein membranes and brings about coalescence. At some point, the oil in water (O/W) emulsion inverts, forming a W/O emulsion. The temperature is well below the melting point of fat, and the solid flakes of butter are suspended in the excess buttermilk. The butter pieces are collected and further processed, e.g., salt may be added. The mass is mechanically worked to improve crystal characteristics, then formed as desired and refrigerated for storage.

If the cream is from fresh milk, the product is called sweet cream butter. Alternatively, the cream may be cultured (inoculated with certain

Rendering, dry—Heating and dehydrating animal tissue to extract and separate the fat. This process results in products with increased color and flavor, so it is not commonly used for edible animal fats.

Rendering, wet—Heating animal tissue by steam under pressure so the fat is released to allow separation.

Ghee—The fat remaining when the water and other components are removed from butter. Also termed anhydrous milk fat, it is nearly 100% fat.

bacteria) and ripened (stored 24–48 hr), which increases the concentration of acids such as lactic acid and numerous flavor compounds (diacetyl and other ketones). Cultured butter is usually unsalted and sold for commercial use, such as a roll-in fat for croissants. Sweet cream butter is generally lightly salted and is preferred for direct consumption.

MARINE OILS

Historically, marine oils meant oil from both fish and whales, but today essentially all marine oil is obtained from fish (menhaden, pilchard, herring, and other species). Marine oils make up about 2% of the annual world consumption of fats and oils and are especially important in countries in northern European, the west coast of South America, and Asia.

Rendering. The fish are netted at sea, refrigerated, and brought to a central processing plant. They are steam-cooked, after which the liquid (oil and water) is removed by hydraulic presses. The water (containing soluble protein) is removed by centrifugation and added back to the press cake. *Dried fish meal* is exceptionally high in protein and is a very valuable animal feed supplement.

Dried fish meal—The material remaining when fish tissue is heated to remove the fat and then dried.

Refining. The crude oil is refined using the same steps as for vegetable oils, including winterizing (see below) to remove triglycerides with high melting points. Marine oils contain a relatively large amount of long-chain (up to C26) polyunsaturated fatty acids. Partial hydrogenation converts them to fats that are useful in the manufacture of margarine and shortening because they strongly stabilize the β' crystal structure.

Hydrogenation—The chemical process of adding hydrogen atoms to the double bonds between carbon atoms in a fatty acid. The result is the conversion of a double bond (unsaturated) to a single bond (saturated).

Manufacture of Food Fats and Oils

With few exceptions (e.g., olive oil) RBD vegetable oils are not used directly in foods. The extent of further processing may involve one to several steps. This processing converts the oils into forms having the required characteristics (clarity, plasticity, oxidative stability) for the intended application.

MODIFICATION OF BASE OILS

Hydrogenation. Hydrogenation was first used industrially in 1897 to hydrogenate unsaturated organic materials and was first applied to vegetable oils in 1903. It converts liquid oils to semisolid plastic fats suitable for manufacturing shortening and margarine. In the process, hydrogen gas is added to the double bonds of fatty acids in the presence of a suitable catalyst (Fig. 4-3). This raises the melting point of the fat and reduces its iodine value (IV). For successful hydrogenation, all impurities must be removed from the oil; gums, soaps, etc. limit the action of the catalyst. By controlling reaction conditions (temperature, hydrogen pressure, and catalyst type and concentration), the processor can

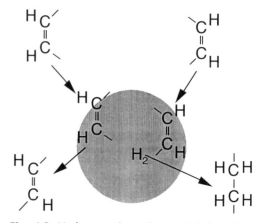

Fig. 4-3. Hydrogenation using a nickel catalyst. The double bond is activated at the surface of the catalyst. Addition of hydrogen gas (H₂) converts the double bond to a saturated bond. If the activated double bond leaves the surface without hydrogenation, cis to trans isomerization occurs, possibly accompanied by migration into a conjugated double-bond configuration.

make end products having greatly varied physical and chemical characteristics.

Hydrogenation forms isomeric unsaturated fatty acids that have properties (especially increased melting point) different from those of the naturally occurring acids. Such isomers originate from the following reactions:

- Addition of hydrogen at a double bond
- Conversion of the natural cis form to the trans form
- Migration of the double bond.

Hydrogenation is performed batchwise in cylindrical vessels made of carbon steel. The height of these reactors is almost twice their width, and they have a high-speed agitator and internal baffles to disperse the hydrogen gas, heating and cooling coils to properly control the reaction temperature, and an inlet port to introduce hydrogen gas. The progress of the reaction is followed by measuring the *refractive index* of oil samples. At the desired end point, the oil is cooled and the catalyst is removed by filtration.

Refractive index—A physical property of a substance that relates to how light is refracted from the material. Usually used to indirectly measure some other property such as concentration.

Selectivity. Selectivity during hydrogenation is based on the fact that the greater the degree of unsaturation of a fatty acid, the greater its chemical reactivity (ability to add hydrogen). Thus, linolenic acid (C18:3) is hydrogenated before linoleic acid (C18:2), which in turn is hydrogenated before oleic acid (C18:1). The relative reaction rates of the usual unsaturated fatty acids are: oleic acid, 1; linoleic acid, 50; isolinoleic (double bonds at the 9,10 and 15,16 positions), 5; and linolenic acid, 100. (Isolinoleic acid is produced by hydrogenation of linolenic acid at the middle [12,13] double bond. Its low reactivity is because the double bonds are isolated.) These relative rates may vary depending upon the particular catalyst being used.

Hydrogenation conditions are said to be selective or nonselective. The parameters that correspond to these conditions are listed in Table 4-1. A single change in any of the process parameters, with the others held constant, affects reaction selectivity, extent of *trans* isomer formation, and reaction rate. These differences, taken together, produce changes in the solid fat index (SFI) profile of the hydrogenated fat. Because SFI is the key to the properties of the margarine or shortening produced from the hydrogenated fat, there is a strong connection between hydrogenation reaction conditions and final product quality.

The amount of unsaturation, as reflected in its IV, does not alone determine the shape of the SFI profile. This is illustrated in Figure 4-4, which shows soy oil samples hydrogenated under selective and nonselective conditions to the same IVs. Nonselective

TABLE 4-1. Hydrogenation Conditions Affecting Selectivity

Reaction Parameter	Selective Hydrogenation	Nonselective Hydrogenation
Temperature	High	Low
H$_2$ pressure	Low (1 atm)	High (>3 atm)
Agitation	Low	High
Catalyst concentration	High (0.05% Ni)	Low (0.02% Ni)
Catalyst type	Selective	Nonselective
Trans-isomer formed	High amount	Low amount
SFI[a] curve shape	Steep	Shallow

[a] Solid fat index.

hydrogenation gives a much flatter profile, while selective hydrogenation produces fat with a humped SFI profile. These differences are used to make partially hydrogenated fats (basestocks), which are blended to achieve an SFI profile appropriate to different applications.

Interesterification. This modification involves moving the fatty acids to different positions in the triglycerides in the fat (see Chapter 1). Random interesterification was initially used to improve the poor creaming properties of lard. Because of the uniform nature of lard triglycerides (all the palmitic acid is in the 2-position), it solidifies in the β phase. By interesterification, the location of this fatty acid is randomized, resulting in the formation of the more desirable β' crystals. This produces a smoother shortening with excellent creaming properties. The rearrangement process does not change the degree of unsaturation or the isomeric state of the fatty acids as they transfer from one position to another. The production of a non-*trans* all-soy oil margarine (discussed in Chapter 1) is another example of this technique.

Directed (or nonrandom) interesterification is done at a temperature below the melting point of the triglycerides to be removed. This permits the rearrangement process and the crystallization of high-melting triglycerides to occur simultaneously, allowing further modification of the fat's physical properties. For example, a fat resembling cocoa butter is made by directed interesterification of hydrogenated cottonseed oil.

The interesterification reaction takes place under vacuum in an agi-

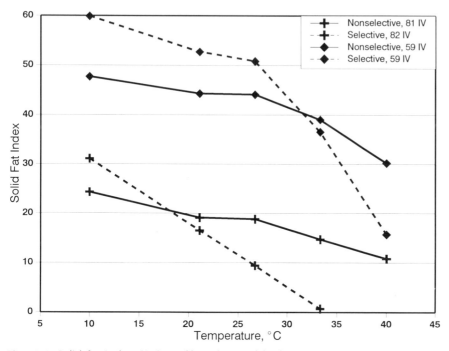

Fig. 4-4. Solid fat index (SFI) profiles of soy oil hydrogenated under selective and nonselective conditions. The selectively hydrogenated samples have a steeper, humped profile (higher SFI at low temperatures, lower SFI at high temperatures) than the nonselectively hydrogenated samples at the same iodine value (IV).

tated stainless steel vessel at an elevated temperature. The reaction is aided by a catalyst (e.g., sodium methoxide), and it is essential that no water be present to form soap, that the fat be free of acids and peroxide, and that the catalyst be rapidly dispersed through the reaction mixture immediately after it is added. Recent work has been directed at using microbial lipase to bring about the migration of the fatty acids. Promising results have been obtained in pilot plant runs, but the use of lipases is not yet economically competitive with the use of inorganic catalysts for full-scale production.

SHORTENINGS

Shortenings are anhydrous fats (unlike margarines, which contain water) used primarily in baking. By proper selection of basestocks, almost any desired SFI profile can be obtained, according to the requirements of the application.

Production of plastic shortenings and table-grade margarines utilizes the same equipment, and the same principles apply to both products, with one exception. Because margarine contains at least 17% aqueous phase, all processing equipment must be constructed of stainless steel, whereas shortening production may utilize heat exchangers constructed of carbon steel. However, all new installations use stainless steel units.

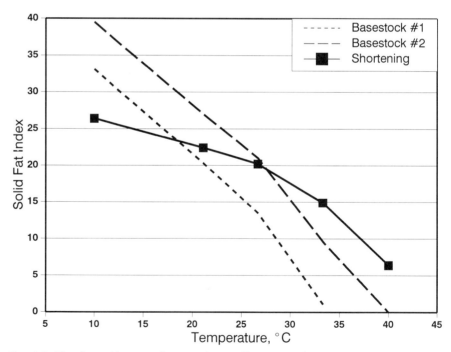

Fig. 4-5. Blending of basestocks to make an all-purpose shortening. Basestock no. 1 was selectively hydrogenated, basestock no. 2 was nonselectively hydrogenated, and the hard flakes were fully hydrogenated (solid fat index >80 at all temperatures). The all-purpose shortening comprised 27% of basestock no. 1, 27% of basestock no. 2, 12% hard flakes, and 34% refined, bleached, deodorized oil.

Modern technology for plasticizing shortening originated about 60 years ago. In 1931, the Girdler Corporation developed the closed, continuous, internally chilled plasticizer, using scraped-wall heat transfer. Since that time, the original unit has undergone several modifications, but the principle has remained the same. Other manufacturers have developed similar systems that embody the same heat transfer concept. The main differences lie in the treatment of the crystallized fat phase after the initial rapid shock-cooling step.

Blending of basestocks. Basestocks are vegetable oils hydrogenated with varying degrees of selectivity, and to different extents, to give fats having SFI profiles ranging from quite steep to rather flat and being higher or lower on the SFI plot. These are blended to make the shortening (or margarine fat) having the desired SFI profile, from the steep profile of a wafer filler fat to the soft, flat profile of a spreadable tub margarine. A manufacturer has a "stable" of such basestocks, from which any shortening in the product line can be made by combining set proportions of certain basestocks. In addition to partially hydrogenated oils, the manufacturer also uses a liquid oil (either RBD or lightly hydrogenated) and a heavily hydrogenated oil (hard flake). An example of basestock blending is shown in Figure 4-5, in which two partially hydrogenated fats were combined with hard flake and RBD oil to make an all-purpose shortening.

The basestocks are blended in the desired proportions, then (usually) deodorized. From the deodorizer, they may be stored under nitrogen gas until needed or may go directly to the crystallizing line to produce shortening. In choosing oil sources for basestocks, it is important to keep in mind the necessity of having β′ crystals in the final product. Most vegetable oils are comprised of mainly C18 fatty acids, and, because of this uniformity, the hydrogenated oils form stable β crystals. Other oils have a greater variety of fatty acid chain lengths, and, when hydrogenated, these form stable β′ crystals. Table 4-2 lists fats in the two categories. If the basestocks are partially hydrogenated soy oils, for instance, the hard flake will usually be 5–10% fully hydrogenated palm or cottonseed.

Some shortenings contain both vegetable and animal fats. In these shortenings, the animal fat (lard or tallow) supplies much of the necessary solid fat phase. Such shortenings are often designated "A/V shortening." They tend to be somewhat less expensive than all-vegetable shortenings but are not Kosher (approved for Jewish and Muslim dietary use).

TABLE 4-2. Crystal Habits of Fats and Oils

Fat Type	Stable in β Form	Stable in β′ Form
Vegetable oil	Canola	Cottonseed
	Coconut	Palm
	Cocoa butter	
	Corn	
	Palm kernel	
	Olive	
	Peanut	
	Safflower	
	Sesame	
	Soybean	
	Sunflower	
Animal fat	Lard	Tallow
		Modified lard
Milk fat		Butterfat
Marine oil		Herring
		Menhaden
		Whale

Votating—A process that mixes, cools, and whips air or other gas into a fat.

Votating. The name "Votator" designates the equipment originally designed by the Girdler Corporation, but the term "votating" has come into common use for the process of converting melted fat into a plastic shortening (or margarine). The closed, continuous Votator system, introduced in 1937, allows chilling and working the product under conditions of maximum sanitation and efficiency. The apparatus consists of a scraped-surface heat exchanger chilling unit, or "A" unit, jacketed for a direct-expansion-refrigerant system such as ammonia or Freon, and a larger "B" unit in which the shortening undergoes further crystallization.

The temperature of the melted oil blend is adjusted in a precooler; then the oil is chilled in the A unit to 15–20°C in 10–20 sec. The high-melting triglycerides begin to crystallize into crystals that quickly transform to needles. Most of the crystallization occurs in the B unit, where low-speed agitators break up the initial crystals, providing seeds for further crystal formation. The temperature of the fat rises 5–10°C due to release of heat from the crystallization. Little or none of the crystal formation should occur in the final package. Nitrogen gas is introduced into the oil as it enters the A unit. A pressure relief valve at the end of the B unit adjusts the gas pressure so that the shortening contains 12–15% N_2 (by volume) as it is extruded into the final packaging. The dispersed gas gives the desired white color as well as providing some protection against exposure to oxygen (air) during storage.

Tempering—During the manufacture of shortenings, holding the package product at carefully controlled temperatures to make slight changes in the crystals of fat. This extends the temperature range over which a shortening remains plastic.

Tempering. After the shortening is packaged, it is moved to a tempering room, where it stands for two to four days at 27–29°C (80–85°F). The phenomenon of tempering is not fully understood, but it probably involves melting and recrystallization of some of the mixed crystals formed during the rapid chilling. A properly tempered fat is slightly firmer above the tempering temperature and slightly softer below that temperature, compared with the same fat that has not been tempered. In practice, this means that the plastic range of the shortening is extended. If the freshly packaged shortening is immediately removed to cold storage, it becomes hard and brittle.

In-plant plasticization. If a facility uses large quantities of plastic shortening (10 tons per week is perhaps the break-even point), it may be economically advantageous to plasticize the shortening in-house. The melted fat blend is delivered in heated tank trucks, pumped to and stored in heated tanks, and processed through a small plasticizer as needed. This has the advantage of requiring less labor in handling. The economies of bulk buying and labor savings must be balanced against the capital and operating costs of the equipment. Several companies supply such equipment and can help with a cost feasibility study.

MARGARINES

Formulations. Standard margarine contains at least 80% fat; the rest is a skim milk, whey, or water solution containing salt and flavors. In preparing margarine, there are six steps:

1. Blending basestocks plus other oil-soluble components

2. Mixing milk or water with salt and other water-soluble ingredients

3. Mixing the two phases to form a W/O emulsion

4. Chilling and plasticizing the above emulsion

5. Forming the margarine into prints or filling it into plastic tubs

6. Finish packaging and cold storage of the finished product.

General Purpose Margarine Specifications				
AOM stability: 200 hr		Solid fat profiles:		
Oil stability index: 80 hr		°C	SFC	°F	SFI
Wiley mp: 47 ± 2°C		10	38 ± 3	50	28 ± 2
Moisture content: 17%		15	30 ± 3	70	21 ± 2
Salt content: 3.0%		20	23 ± 2	80	18 ± 1
		25	17 ± 1	92	15 ± 1
		30	13 ± 1	104	10 ± 1
		35	10 ± 1		
		40	7 ± 1		
		45	4 ± 1		

In step 1, the oils to be used are weighed into a tank (at 5°C above their melting points) and emulsifier (lecithin, mono- and diglyceride, or a combination), oil-soluble vitamins (A, D, and E), and colorants (β-carotene, annato) are added. The oil is then mixed, preferably under a nitrogen atmosphere to prevent oxidation and flavor deterioration.

When the oil-soluble ingredients are thoroughly mixed, the water phase is added. This could be whole milk, skim milk, reconstituted nonfat dry milk solids, or water, plus salt. Antimicrobial agents (sodium benzoate, potassium sorbate) and/or the heavy metal-chelating agents citric acid and ethylenediaminetetraacetic acid are added if needed and if the law allows. Flavors such as diacetyl and starter distillate are also included in the water phase. This phase is slowly added to the oil with agitation to form a W/O emulsion. Sometimes the emulsion is pasteurized at 73°C (165°F) for 16 sec before being chilled in the Votator, to ensure freedom from pathogenic bacteria.

Votating. The margarine oils are chilled and plasticized as described above for shortening manufacture. Margarine is a stable emulsion because the solidified oil contains the dispersed aqueous phase in droplets ranging in diameter from 2 to 20 μm, with the majority about 5 μm. In the Votator process, the violent agitation and kneading in the A unit produce this extremely fine dispersion of the water phase droplets in the fat phase. The B units in a margarine line are quiescent; they do not have an internal agitator. The margarine is allowed to solidify for a time before being extruded to the packaging line. One other difference from shortening manufacture is that nitrogen gas is not introduced into a margarine line; the emulsified water droplets provide the desired opacity.

SALAD OILS

Winterization. Originally this process was applied to cottonseed oil intended for use as

Specifications Common to All Good-Quality Fats and Oils
Chemical characteristics
Peroxide value: 1 meq/kg maximum
Free fatty acid (as oleic acid): 0.05% maximum
Phosphorus content: 1 ppm maximum
Physical characteristics
Color (Lovibond): 1.5 R, 15 Y maximum
Flavor: Bland
Odor: Neutral when warmed

Winterization—A process in which salad oils are cooled until high-melting-point triglycerides form crystals. These crystals are removed so that the next time the oil is cooled, the cloudiness that comes with crystallization will not occur.

salad oil or for making mayonnaise. It removes the high-melting triglycerides that would otherwise precipitate at refrigerator temperatures, breaking the mayonnaise emulsion. The oil was subjected to slow crystallization in outside tanks during winter (hence the term "winterization"), and the solid fat was filtered off. Today, mechanical refrigeration is used to winterize cottonseed, lightly hydrogenated soybean oil, and a host of other oils. The properties of the solid fat portion depend on the temperature and time of crystallization. This portion is used in other food applications, for example, as shortening hard flakes, for frying fats, and as confectionery fats.

FRYING OILS

The subject of frying fats is discussed in detail in Chapter 6. This chapter briefly outlines the processing for producing different kinds of frying fats.

Consumer grade. Most home deep-frying is done with salad oil from the supermarket shelf. Because these are usually just winterized RBD oils, their frying stability is not high, and they are not used many times or held at high temperatures. A good grade of salad oil suffices, although peanut or cottonseed oil is generally considered preferable to soy or sunflower oil for this use.

High-stability commercial fats. These may be either plastic fats or liquid oils, depending upon the desired characteristics in the finished product. Frying fat has a higher SFI profile than most shortenings (see Appendix C), but its melting point is not much higher than body temperature. These characteristics are obtained by selective hydrogenation of the base oil.

Frying oils are made by hydrogenating oil to reduce most of the polyunsaturated fatty acids to monounsaturated acids. After hydrogenation, the oil is fractionated to remove the high-melting triglycerides. The oil remaining has a melting point around room temperature and high oxidative stability; one commercial product specifies that its oxidative stability by the active oxygen method is 350 hr minimum. In dry fractionation, the hydrogenated fat is melted, then cooled under controlled conditions. The oil and solid phases are separated by filtration. In solvent fractionation, a solvent (e.g., hexane or acetone) is added to the warm fat. This expedites crystallization and subsequent filtration. Of course, the solvent is stripped from both phases after separation.

Troubleshooting

Plastic Fats

Symptom	Causes	Changes to Make
Oil on carton	Storage at high temperature	Keep warehouse below 35°C (95°F).
Grainy shortening	Crystal formation	Store below 35°C. Use some β'-tending fat in the oil formulation.
Difficulty in creaming or in rolling-in the fat	Fat too hard (or soft) for proper functioning	Hold fat at application temperature (about 25°C [77°F]) for 24 hr before use. Make sure SFI[a] specification suits the application.
Dough tears during roll-in	Fat is too hard at low temperatures	Decrease SFI specification at low temperature. Make sure shortening was properly tempered.
Fat oils out during roll-in	Fat is too soft at high temperatures	Increase SFI specification at high temperature. Make sure shortening was properly tempered.
Improper color	Yellowish cast	Inadequate bleaching by manufacturer. Partial transformation to β phase.
	"Vaseline" streaks in the shortening	Improve N_2 dispersion during manufacture (probably does not affect functionality).

Bulk Oils

Symptom	Causes	Changes to Make
Rancid, reverted flavor	Autoxidation, high peroxide value	Manufacturer must perform proper deodorization. Fill headspace in all storage tanks with N_2 gas.
Rapid development of rancid flavor in product	High peroxide value when oil was used	Ensure proper bleaching and deodorization by supplier. Store oil under N_2 gas. Don't allow contact of oil with metal (iron, copper). Include chelators (citric acid) and antioxidants in oil.

In-Plant Oil Plasticization

Symptom	Causes	Changes to Make
Fat from plasticizer is too soft	Fat temperature too high	Cool oil before entering plasticizer. Increase capacity of the cooling unit of plasticizer.
	Not enough crystals formed during plasticization	Increase residence time in crystallizer. Specify higher SFI profile for oil.
	Inadequate immobilization of the oil phase	Increase amount of stable source oil.
Fat from plasticizer is too hard	Fat temperature too low	Decrease amount of cooling applied to crystallizer. Specify lower SFI profile for oil.
Fat consistency varies during run	Solids in storage tank are settling	Raise storage temperature to the complete melting point. Install adequate stirrer in tank. (Make sure it does not whip air into the oil!)

Bulk System Sanitation		
Symptom	Causes	Changes to Make
Soapy flavors	Water in the system	Ensure complete drying of system after cleaning.
Sporadic rancid flavors	Old oil entering the main stream	Make sure there are no dead spots in the system. Ensure that the clean in place (CIP) system removes all oil during operation.

[a] Solid fat index.

Bakery Product Applications

Functions in Baked Goods

Fats and oils have been important bakery ingredients for centuries. Indeed, "shortening" is a baker's term; fat in a bakery item "shortens" (tenderizes) the texture of the finished product. Shortening also plays an important part in the efficient processing of many bakery products. The primary properties of a fat or oil that determine its ability to carry out these functions are:

- Its ratio of solid to liquid phase
- Its oxidative stability
- The *plasticity* of a solid shortening.

The physical basis of these properties has been discussed in Chapter 1. This chapter discusses how they interact with other aspects of baking (e.g., ingredients, processing) to produce the desired final result.

Emulsifiers also influence how shortening functions. Thus, any discussion of the role of fats and oils in bakery products must include emulsifiers and how they affect final product quality.

PRODUCT CHARACTERISTICS

The contributions that shortening makes to the final quality of bakery products are: tenderness, moist mouthfeel, *lubricity*, flavor, and structure. Here are some examples.

Tenderness. A basic white bread, formulated without shortening, is somewhat tough to the bite. The inclusion of 3% shortening (either fat or oil) in the formula gives a *crumb* that is softer and easier to chew.

Moist mouthfeel. A layer cake made with oil as the shortening gives the impression of a moister crumb than does cake from the same formula made with an emulsified plastic shortening. This difference becomes more noticeable after storage for a few days.

Lubricity. A product with lubricity clears readily from the surfaces of the mouth. If you eat one snack cracker directly after it exits the oven and one after it has passed through the oil sprayer and then compare their eating quality, the lubricating effect of the spray oil is apparent. The unsprayed cracker is "dry" and difficult to clear from the palate, whereas the sprayed cracker is more readily chewed and swallowed.

In This Chapter

Functions in Baked Goods
  Product Characteristics
  Processing

Hard Wheat Products
  Bread, Rolls
  Laminated Doughs
  Yeast-Raised Doughnuts

Soft Wheat Products
  Cake
  Cake Doughnuts
  Cookies
  Crackers
  Pie Crusts, Biscuits
  Filler Fats and Icings

Pan-Release Agents

Troubleshooting

Plasticity—A physical property of a fat that describes how soft, pliable, and moldable it is at a given temperature.

Lubricity—A desirable slippery sensation in the mouth imparted by fats.

Crumb—The interior of a baked product as distinct from the crust.

Keyholing—A process in which weak side walls of a bread loaf collapse inward so that the final product is shaped like a keyhole.

Continuous bread process—A method of dough formation and bread preparation in which ingredients are added, mixed, and fermented without stopping points in the process.

Hard flake—A hard fat hydrogenated to an iodine value of about 5 and having a melting point of about 57°C (135°F).

Sweet goods—Baked products such as Danish pastries that are high in sugar and often have added icing.

Extrusion—The process of putting a product under pressure and sometimes heat and then forcing it out of a barrel through an orifice.

Wire-cut cookies—A type of cookie prepared from a dough in which the individual cookies are cut from the dough piece with a wire before baking.

Flour basis—When determining the percentage of an ingredient in a formula, the weight is compared to the weight of the flour as a percentage. A formula with 50 lb of flour and 5 lb of fat would contain 10% fat on a flour basis.

Oven-spring—The increase in volume that occurs when baked products are first put into the oven. It results from the release or expansion of gases that takes place before the structure of the product is set.

Flavor. Fat contributes flavor to many baked products. This is especially apparent in fried products such as doughnuts. For example, although a yeast-raised doughnut or bench-cut cake doughnut could be baked in an oven, the flavor is less appealing than the same doughnut that has been deep-fried. Another way to demonstrate the same effect is to fry some cake doughnuts in regular frying fat and some (from the same batter) in food-grade mineral oil. Panelists who sample the doughnut fried in mineral oil report that it has "no flavor," as compared to the regular product. The flavor imparted by the fat is attributable to the many products formed by reactions of the triglycerides and fatty acids with proteins, sugars, and air. At high concentrations, these flavor compounds become objectionable (when the fat has been abused), but at the trace levels found in properly treated frying fat they give the appetizing flavor notes we associate with deep-fried foods.

Structure. The solid fat portion of plastic shortening (e.g., all-purpose shortening) plays a structural role in some products. It strengthens the sidewalls of bread, minimizing "*keyholing*" of the finished loaf. In the *continuous bread process*, 5–10% *hard flake* is often melted into the warm oil being pumped to the developer to provide this strength. Hard flakes are frequently incorporated into icings (such as *sweet goods* icings) during preparation. Gums are used to stabilize the interaction between water and sugar in the icing, and the fat gives the icing flexibility so that it does not crack and flake off the product. In a creme icing, the solid fraction of butter or an emulsified icing shortening stabilizes the icing, as well as entrapping air during the whipping step.

PROCESSING

Shortening influences the processing steps of many bakery products. Some of the factors it affects include: air incorporation, structure during processing, lubrication, and heat transfer.

The solid portion of a shortening entraps air bubbles during mixing. This is particularly important in products such as cakes and cookies, as explained later in the section on soft wheat products. The solid fat also contributes to proper consistency during the *extrusion* of icings and fillings.

The oil phase in a shortening lubricates dough during extrusion processes, such as in a *wire-cut cookie* extruder. Oil also aids the release of dough from rotary cookie molds and finished products from pans. Oil is an efficient heat exchange medium during deep frying.

Hard Wheat Products

BREAD, ROLLS

Volume. The inclusion of shortening in a bread formula increases the volume of the baked bread (Fig. 5-1). The maximum volume is obtained with approximately 6% (*flour basis*) shortening, but in practice 3–5% is generally used. The larger volume is the result of an increase in *oven-spring*. The loaf expands as its internal temperature rises during baking.

Then, over a temperature range of about 60–70°C (140–158°F), the starch granules gelatinize. At a somewhat higher temperature (80–90°C, 176–194°F), gluten and other proteins denature. These two reactions stop the expansion. Fat seems to interact with dough components (starch and gluten) to delay the reactions that end loaf expansion during baking. In other words, shortening allows the loaf to expand longer before setting, giving a larger final volume. The exact mechanism of this effect is not clear, but various studies confirm that 5% shortening increases loaf volume by 15–25% as compared to a control containing no shortening.

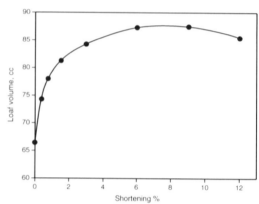

Fig. 5-1. Effect of shortening on volume of white bread. The individual loaves each contained 10 g of flour ("pup" loaves).

Softness (antistaling). Staleness of bread is a matter of degree; there is no sharp line between "fresh" and "stale" bread. Staling of bread is usually evaluated by measuring the firmness of the bread crumb. In the laboratory, this is done with an instrument such as the penetrometer or the Instron Universal Testing Machine; the consumer simply squeezes the loaf. Stale bread tastes dry, feels harsh, and lacks flavor. Estimates of staleness by sensory evaluation correlate highly with simple firmness measurements.

Staling results mainly from the recrystallization of amylopectin. During baking, starch granules gelatinize; the linear amylose molecules escape (in part) from the granules, and the branched amyopectin molecules expand (but they remain within the swollen granules). During cooling, the solubilized amylose recrystallizes, allowing the bread crumb to become firm enough for slicing. The amylopectin branches recrystallize over a period of several days, causing the granules to harden and leading to crumb firmness (Fig. 5-2).

Shortening reduces the firmness of the bread throughout its storage life. By the fifth day of storage, bread containing shortening has the same firmness as the control bread during the third day. In other words, its acceptability is extended by 1–2 days (Fig. 5-3). Plastic shortening contains approximately 25% solids at room temperature, so 3% refined, bleached, deodorized (RBD) oil gives the same softening effect as 4% plastic shortening.

Certain emulsifiers also slow the staling rate. Monoglycerides of saturated fatty acids (for example,

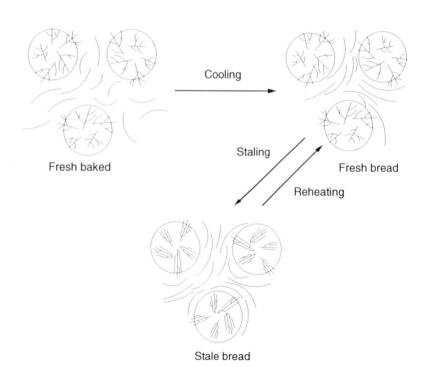

Fig. 5-2. Changes in starch molecules in bread. Immediately out of the oven, both amylose (straight chains) and amylopectin (branched structures) are swollen and randomly oriented. During cooling (1 hr), the amylose molecules begin to align and crystallize. During staling (several days), amylopectin realigns and reforms crystallites.

monostearin) are the most effective, but many other emulsifiers that contain saturated fatty acids (e.g., stearoyl lactylate) also have some activity. The linear hydrocarbon part of the fatty acid chain forms a complex with the helical starch molecules. When this complex involves the branched chains of amylopectin, it interferes with amylopectin recrystallization, thus slowing the development of firmness (Fig. 5-3). This interaction was discussed more fully in Chapter 3.

The properties of the protein (i.e., denatured gluten) in bread may also alter during storage and possibly contribute to staling. While some research has been done in this area, the role of protein in staling is not clear at this time. More work may lead to a better understanding of how some emulsifiers might retard staling by interacting with the protein in baked bread.

Crumb strength. The solid fat portion of traditional bread shortening (e.g., lard or all-purpose shortening) provides dough strength during *proofing* and resistance to shocks during transfer from the proofer to the oven. It also strengthens the sidewalls of baked bread, minimizing "keyholing" of the finished loaf. When RBD oil is used to replace plastic shortening, this strengthening effect is maintained by the addition of sodium stearoyl lactylate or DATEM esters to the dough, up to the legal maximum of 0.5% (flour basis). In the continuous bread process, hard flake is melted into the warm oil being pumped to the developer to provide strength.

> Shortening in bread slows staling and increases the bread's shelf life.

Proofing—A step in preparing yeast-leavened products in which the dough is warmed and allowed to rise. It takes place after an initial fermentation and before baking.

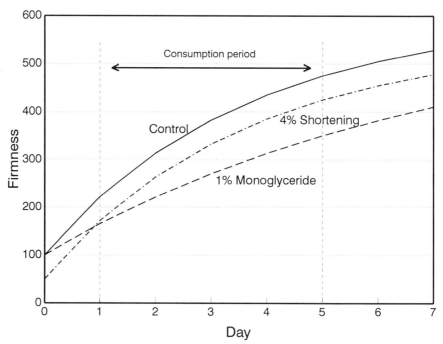

Fig. 5-3. Effect of shortening and monoglyceride on bread softness during storage. Shortening decreases firmness by a constant amount throughout storage. Monoglyceride doesn't change initial firmness, but it slows the rate at which crumb firmness develops.

LAMINATED DOUGHS

Laminated doughs are the basis for numerous specialty bakery products, from Danish to Napoleons, from croissants to phyllo dough. The base dough may be either rather rich and sweet or lean. After *development*, the dough is divided into pieces (weighing 10–16 lb) and *retarded* (held at 5–7°C, or 40–45°F) for 2–4 hr. The dough is then rolled out into a rectangular shape; shortening is spread over two-thirds of the dough; and the dough is folded in thirds to give three layers of dough separated by two layers of fat (called a "book"). The amount of shortening ranges from 17 to 36% of the dough weight, depending upon economics and the desired product quality. The book is then rolled out again into a rectangle. It is folded in thirds, and the process is repeated until the desired final number of laminations is reached. The rolling process warms the dough, and usually the book is returned to the retarder partway through the process for cooling.

Equipment has been developed to perform this process automatically. The shortening is extruded onto a continuous dough sheet or co-extruded with dough, and then various mechanical methods are used to obtain the desired number of laminations. Just as with the manual method, the proper selection of fat is important to obtain a good quality laminated dough.

Danish. Roll-in shortening or margarine for laminated doughs must have a broad plastic range. Fat consistency should match dough consistency at retarder temperatures and at room temperature or above. If the fat is significantly harder than the dough at the cool temperature, then when a retarded dough is rolled out, the fat is likely to tear holes in the dough sheet. If the fat is softer than the dough at room temperature, then as the dough mass warms up (during laminating) the shortening soaks into the dough, the adjacent dough layers knit together, and the layering effect is lost.

The proper plasticity of roll-in fat requires a relatively shallow SFI profile and stabilization in the β′ crystal phase. The melting point of the fat must be higher than the proofing temperature. If the proof box temperature is above the fat melting point, the fat layers turn to oil. This allows the dough layers to knit together during proofing, and the final product is less flaky than desired. As a general rule, the complete melting point should be about 5 degrees C (9 degrees F) higher than proof box temperatures.

Puff pastry. The formation of large voids between dough layers in puff pastry products is due mainly to the evaporation of water in the roll-in margarine. The margarine is divided into small pieces, thinly *sheeted* between dough layers during roll-in. In contrast to Danish dough, a continuous fat layer is not desired, so initially

Laminated dough—A dough system that has horizontal layers of dough separated by layers of fat, resulting in a flaky baked product.

Dough development—The process of forming gluten from wheat flour with the addition of water and mixing.

Retardation of dough—A cooling step that slows down the development of dough and the growth of yeast and allows fats to become more solid.

Specifications Common to All Good-Quality Fats and Oils

Chemical characteristics
  Peroxide value: 1 meq/kg maximum
  Free fatty acid (as oleic acid): 0.05% maximum
  Phosphorus content: 1 ppm maximum
Physical characteristics
  Color (Lovibond): 1.5 R, 15 Y maximum
  Flavor: Bland
  Odor: Neutral when warmed

Puff Paste Margarine Specifications

AOM stability: 200 hr
Oil stability index: 80 hr
Wiley mp: 54 ± 2°C
Moisture content: 18%
Salt content: 2%

Solid fat profiles:

°C	SFC	°F	SFI
10	47 ± 4	50	34 ± 3
15	40 ± 4	70	30 ± 3
20	33 ± 3	80	27 ± 2
25	27 ± 2	92	22 ± 2
30	21 ± 2	104	16 ± 1
35	16 ± 1		
40	11 ± 1		
45	7 ± 1		

Sheeting—Stretching dough horizontally to develop the gluten structure.

Baker's margarine—A product similar in composition to butter but containing hydrogenated vegetable oil rather than butterfat.

> For laminated doughs, the complete melting point should be 5 degrees C (9 degrees F) higher than the proofing temperature.

the margarine is spotted on the rectangle of dough in individual pieces rather than in a continuous layer. The individual fat pieces must maintain their integrity through proofing, until the pastry is baked. Thus the SFI of puff pastry margarine is higher than that of roll-in margarine. The "book" is usually rolled and folded twice (rather than three times as for Danish) giving about 18 laminations. This produces the flakiness and open texture characteristic of puff pastry.

Croissants. The traditional roll-in fat for croissants is unsalted butter, and this is still used in the highest quality product. Butter contributes both flavor and color to the final product. Because of its rather steep SFI curve, the temperature range for rolling-in is rather narrow, about 18–20°C (65–68°F). This may necessitate retarding between each rolling step, depending upon the amount of heat generated by working the dough. Proofing temperatures must not exceed the melting point of butter (approximately 36°C, or 97°F). *Baker's margarine* works well during roll-in, but the flavor of the baked croissant is inferior to that of butter croissants. Puff pastry margarine or shortening can be used but leaves a "waxy" taste because of its high SFI at body temperature.

YEAST-RAISED DOUGHNUTS

Shelf life. Yeast-raised doughnuts are fried in hot fat. During frying, a certain amount of fat is absorbed into the doughnut. This fat softens the doughnut crumb and extends the shelf life of the finished product. The keeping quality of the doughnut is directly related to its fat content, up to the point at which the doughnut has a noticeably "greasy" texture and flavor. The factors that govern extent of fat absorption are similar to those for cake doughnuts, which are discussed in more detail below.

Soft Wheat Products

CAKE

> Creating many small air bubbles in a cake batter produces high product volume and fine crumb grain.

Air incorporation. In layer cakes and related items (muffins, cake doughnuts), the closeness of the internal grain, and to some extent the final volume, are strongly influenced by the characteristics of the shortening used. In the finished cake, a high percentage of the total volume is open space, present as finally divided cells. These spaces are created by carbon dioxide (from the leavening system) and steam (formed during baking). When these gases are generated by heat, they migrate to the nearest air bubbles, which were incorporated into the cake batter during mixing. If there are many small air bubbles in the batter, the leavening gases are distributed widely. Each of the bubbles is small and so does not rise rapidly to the surface of the cake. The leavening gases are retained in the cake and contribute to final volume. However, if the air incorporated during mixing is present as a relatively few large bubbles, then during baking when the bubbles are expanded by the leavening gases, many of them are large enough to rise to the surface, escaping and yielding a lower final cake volume. Those that don't escape may form tunnels in the cake. In sum, if the original batter contains many small air cells, the final cake will have larger volume and a fine (close) grain. If the original air bubbles are fewer and larger, the final cake will have less volume and a coarse (open) grain. The shortening used plays a large role in determining the degree of subdivision of the air.

In the usual (retail bakery) three-stage process of cake production, the shortening and sugar are combined and mixed. During this step, the air is dispersed in the solid phase. Then the eggs are incorporated, followed by the flour, liquids, and other ingredients. During the first, *creaming* step, the plastic shortening entraps air bubbles. In the presence of an emulsifier such as monoglyceride, these bubbles are divided into numerous small air cells by the action of the beater. The shortening must be solid so that the bubbles don't escape but also plastic so that it can fold around each air pocket. This is best accomplished by a plastic shortening crystallized in the β′ phase. If the shortening has transformed into the β phase, the large plates of solid fat are much less effective in entrapping the air. A good shortening for this type of cake batter production has the SFI profile of all-purpose shortening, containing about 4% monoglyceride.

> **Creaming**—High-speed mixing of a plastic shortening containing sugar in order to incorporate air.

The influence of emulsifier in promoting air dispersion, and the effect on final cake characteristics, is demonstrated in Figure 5-4. The emulsified shortening contained monoglyceride at 4.5%, a typical level for cake shortening. Other emulsifiers also give good results. Thus, sucrose esters have shown functionality in some tests. An emulsifier system (13% each, by weight, of Polysorbate 60, sorbitan monostearate, and plastic mono- and diglycerides, hydrated in 61% water) used at 2% of the flour weight along with an unemulsified plastic shortening gives better volume and finer grain than a control cake made with high-ratio emulsified shortening.

It is possible to make cakes in a one-stage production process, in which all the ingredients are added at once and the batter is mixed in a single step. In this case, the air is entrapped in the water phase rather than in the shortening. To form this air-in-water foam, which is then stabilized by proteins contributed from flour and eggs, it is necessary to prevent the defoaming action usually associated with fats and oils. This is accomplished by including an α-tending emulsifier in the shortening. Typical ones used are propylene glycol monoesters (PGME), acetylated mono-

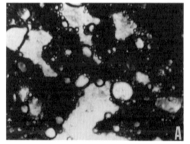 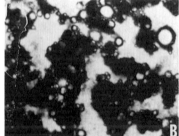

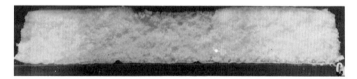

Fig. 5-4. Photomicrographs of cake batters. A, Plastic shortening, no emulsifier, batter specific gravity 0.85 g/ml. B, Plastic shortening, 4.5% monoglycerides, batter specific gravity 0.81 g/ml. The dark areas in A and B show finely dispersed air bubbles contained in the fat phase. C and D, Cakes baked from A and B, respectively.

AOM stability: 75 hr

Oil stability index: 30 hr

Wiley mp: 46 ± 1°C

α-Monoglycerides: 3.5% for
cake; 2.5% for icing

Solid fat profiles:			
°C	SFC	°F	SFI
10	44 ± 4	50	32 ± 3
15	35 ± 3	70	25 ± 2
20	28 ± 3	80	22 ± 1
25	22 ± 2	92	16 ± 1
30	17 ± 2	104	11 ± 1
35	11 ± 1		
40	7 ± 1		

glyceride (AcMG), or lactylated mono-glyceride (LacMG), at a concentration of 7–10% of the plastic shortening. A number of other emulsifiers have also been used, including Polysorbate 60, stearoyl lactylic acid, and sucrose esters, but most commercial cake emulsifiers (and consumer cake mixes based on these systems) utilize PGME and AcMG.

Cakes made with oil as the shortening are more tender than those made with a plastic shortening. The cake gives an impression of moistness when eaten, even after storage for a week or longer. In commercial cake production, unfortunately, oil cakes are often too tender to withstand the necessary handling along the production line. However, tenderness is highly desirable from the consumer's standpoint, and so packaged cake mixes for home use are usually made with oil as the shortening. During the preparation of cake batter from such a mix, the air is incorporated as a foam, which is stabilized by the soluble proteins that are present (i.e., egg white, soluble flour protein). However, as is well known, oil destabilizes protein foams. (For instance, egg whites cannot be whipped into a meringue if there are traces of oil present in the bowl or on the beaters.) This problem is overcome by the use of α-tending emulsifiers.

Fig 5-5. The solid film of α-tending emulsifier left after some water was withdrawn from a water droplet suspended in cottonseed oil containing propylene glycol monostearate (PGMS). The PGMS had formed this film at the interface between the water and the oil.

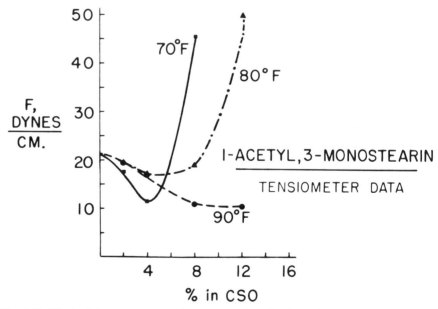

Fig 5-6. Effect of temperature and concentration (in the oil phase) of 1-acetyl-3-monostearin on the formation of a film at the oil-water interface. The interfacial tension was measured with a Tensiometer, which measures the force required to pull a ring through the interface. The apparent increase in interfacial tension at higher emulsifier concentrations is actually due to the presence of a solid film of emulsifier. CSO = cottonseed oil.

These emulsifiers form a solid film at the oil-water interface (Fig. 5-5) when dissolved in the oil above a certain concentration. This solid interfacial film segregates the oil from the water phase, so it cannot destabilize the protein foam. A definite relationship exists between temperature and the minimum bulk concentration of the emulsifier necessary to produce an interfacial film (Fig. 5-6). The addition of a second surfactant may enhance film formation; a mixture of propylene glycol monostearate (PGMS) and stearic acid (80:20) is a stronger film former than pure PGMS at the same total weight concentration. A common system for oil cakes is 10–14% PGME plus 1–2% stearic acid dissolved in the oil.

CAKE DOUGHNUTS

Air incorporation. In a cake doughnut batter, the degree of subdivision of air entrapped in the shortening phase has much to do with the openness of grain in the final doughnut. Thus, for the same reasons as discussed above, an emulsified cake shortening is generally used. Lecithin is also frequently incorporated into the dry mix formula at 0.25–0.5% of total mix weight. This aids in wetting the dry mix during batter preparation.

Frying and fat absorption. About half the fat in the fried doughnut is absorbed from the fryer during frying. In a typical situation, the weight of fat absorbed is approximately equal to the weight of water lost to evaporation. This is explained in more detail in Chapter 6. The degree of fat absorption is influenced by emulsification during frying. For example, emulsifiers such as Polysorbate 60 or ethoxylated monoglyceride help give a finer grain in the finished doughnut, but they also expedite the evaporation of water and pickup of fat. The free fatty acid (FFA) in the frying fat also is an emulsifier. In absolutely fresh fat, with an FFA of less than 0.1%, fat absorption is minimal, and the resulting product is less tender and lacks flavor. As FFA increases (due to hydrolysis) during repeated uses of the fryer, absorption increases. Usually an FFA content of 0.25–0.75% in the frying fat is considered optimal. In a continuously operated system, where fresh fat is added to replace that removed by absorption, this level of FFA is readily maintained. If frying is intermittent, however, the FFA level may exceed 1%, producing excessive fat absorption. Ingredients that hinder water evaporation (e.g., proteins and water-binding fibers) lower fat absorption by the doughnut.

The temperature of the frying fat influences crust color and fat absorption. Generally a temperature of 188–193°C (370–380°F) is recommended. A lower temperature results in a pale crust and excessive fat pickup. A higher temperature gives a dark crust (and possibly incomplete internal cooking, if the frying time is shortened) and also hastens frying fat degradation, which can result in higher fat absorption. Good temperature control on the fryer is imperative.

The physical characteristics of the frying fat markedly influence the eating characteristics of the finished doughnut. If the fat has a rather high SFI curve and a high melting point, the cooled doughnut has a solid texture and a dry mouthfeel. On the other hand, if a high-stability frying oil is used (with a melting point at or below room temperature), the doughnut will be greasy and unappetizing. A good doughnut frying fat has an SFI somewhat higher than that of an all-purpose shortening, but it has a lower melting point (see Table 5-1). When the doughnut is to be coated with powdered sugar, the consistency of the absorbed fat is important in obtaining proper coverage. If the fat in the cooled doughnut is too hard, sugar pickup is too low. On the other hand, if the fat is too soft, the sugar will "melt" when the packaged doughnut is stored. For best results, in warm weather the SFI curve of the frying fat should be on the upper side of the specifications given in Table 5-1, while in cold weather it should be on the lower side.

Doughnut sugar. Sugar for coating doughnuts contains powdered dextrose (corn sugar, glucose) for sweetness and an initial "cool" mouthfeel, starch as a flow agent, and 5-8% fat to provide adhesion of sugar to the doughnut. The fat used has a rather flat SFI curve, as shown in Table 5-1. During the summer, the shortening SFI should be on the high side of this specification, while during the winter, it should be on the low side.

TABLE 5-1. Typical Physical Properties of Bakery Shortenings

| Shortening Type | Solid Fat Index | | | | | AOM[a] (hr, minimum) | OSI[b] (hr, minimum) | Melting Point | |
	50°F 10°C	70°F 21°C	80°F 27°C	92°F 33°C	104°F 40°C			°C	°F
RBD oil[c]	0	…	…	…	…	10	4	<0	<32
Lightly hydrogenated oil	<5	<1.5	…	…	…	25	10	15	60
Deep frying fat	47 ± 3	32 ± 3	25 ± 2	12 ± 1	<2	200	80	36	97
Doughnut frying fat	34 ± 3	24 ± 2	21 ± 2	17 ± 1	9 ± 1	100	40	48	118
All-purpose	28 ± 3	20 ± 2	17 ± 1	13 ± 1	7 ± 1	75	30	47	117
Cake, icing[d]	32 ± 3	25 ± 2	22 ± 2	16 ± 1	11 ± 1	75	30	48	119
Pie crust	26 ± 3	17 ± 1	14 ± 1	10 ± 1	6 ± 1	75	30	45	113
Wafer filler fat[e]	55 ± 4	39 ± 3	29 ± 3	4 ± 1	<1	100	40	36	97
Sandwich cookie filler	38 ± 3	24 ± 2	18 ± 1	9 ± 1	<2	100	40	38	102
Donut sugar fat	18 ± 2	15 ± 2	12 ± 2	10 ± 2	5 ± 2	…	…	45	113
Confectionery coating fat	64 ± 5	52 ± 4	44 ± 4	20 ± 2	0	200	80	37	99
Cocoa butter	75 ± 5	72 ± 5	62 ± 5	3 ± 1	0	…	…	34	94
92 Coconut spray oil	59 ± 5	33 ± 3	6 ± 1	1 ± 1	0	100	40	33	92
Puff pastry margarine	34 ± 3	30 ± 3	27 ± 2	22 ± 2	16 ± 1	200	80	54	130
Baker's margarine	28 ± 2	21 ± 2	18 ± 1	15 ± 1	10 ± 1	200	80	47	117
Butter	32 ± 2	12 ± 1	9 ± 1	3 ± 1	0	…	…	36	97

[a] AOM = Active oxygen method.
[b] OSI = Oil stability index.
[c] RBD = Refined, bleached and deodorized oil.
[d] Cake contains 3.5–5% α-monoglyceride; icing contains 2–3% α-monoglyceride.
[e] Examples only. Variations depend on temperature and other conditions.

COOKIES

Air incorporation and cookie spread. Air bubbles entrapped in the shortening phase of a cookie dough serve as nuclei for leavening gas during baking, as in cakes. Thus, if a fine-grained cookie (for example, a soft sugar cookie) is desired, an emulsified shortening should be used. For a thin, crisp cookie, a lightly hydrogenated oil may give the best product characteristics. For most wire-cut cookies, an all-purpose shortening works well.

The amount of shortening influences cookie spread, but in a way that is related to the sugar content of the dough. Detailed studies, using a sugar cookie test formula, lead to the following generalizations:

• When sugar is used at 50% of the flour weight, increasing shortening from 35 to 55% (flour basis) increases spread about 25%.

• When sugar is used at 90% of the flour weight, increasing shortening from 35 to 55% decreases spread about 10%.

Other formula variations (e.g., the addition of whole egg) also change the amount of influence of shortening on spread, as does the use of different kinds of fats. Thus the safest statement is that each situation is different, and the best way to evaluate the effect of fat level on spread is by experimentation.

Lubrication during processing. The shortening in *rotary-molded cookie* dough is an important processing factor. The oil portion must lubricate the surface of the piece so that it releases cleanly from the mold, while the solid fraction helps maintain shape definition of the piece during molding and baking. To obtain the precise balance required, a combination of lightly hydrogenated oil plus all-purpose shortening in the dough may be necessary. The exact ratio for best results is determined by experiment. If bits of dough stick in the mold, use a little more oil, and if the pattern tends to smear out during baking, use less oil (or a plastic shortening with a higher SFI profile).

In low-fat cookies, difficulty is often seen at the wire-cut depositor; the dough does not extrude smoothly. Using oil as the added allowable lipid may help solve this problem. In these products, the incorporation of air bubbles in the dough is stabilized by soluble proteins and gums.

CRACKERS

Cracker formulas vary quite widely, with a range of 8–12% shortening (flour basis) in the dough. The fat used is typically an all-purpose plastic shortening, which serves to lubricate the gluten sheet, reducing horizontal shrinkage when the dough is sheeted and cut.

Spray oil. Because crackers are baked to a very low moisture content of about 2%, they give a dry mouthfeel when eaten. In the case of snack crackers, this is undesirable. This type of cracker is typically sprayed with an oil while hot, to give a product that clears the mouth surfaces more readily. The usual application level of spray is 15–17% oil, relative to the unsprayed cracker.

> Cookie spread is influenced by the interaction of shortening and sugar.

> **Rotary-molded cookies—** Cookies formed by pressing dough into a mold engraved on a cylinder.

92° Coconut oil—Coconut oil that has been partially hydrogenated to attain a melting point of 92°F (33°C).

The traditional spray is *92° coconut oil*. In recent years, selectively hydrogenated soy oil has been successfully used as a replacement. The polyunsaturated fatty acids are reduced to almost zero, but the melting point is around 95°F (35°C). If a shiny appearance is desired, an oil with an SFI around 10 at 70°F (21°C) is used; oil with an SFI of about 20 at 70°F gives a drier appearance. Oxidation stability is important for spray oils, and AOM values must be at least 100 hr. If an oil is used that has a lower AOM, oxidative rancidity develops during storage, which is detected as a musty odor when the package is opened. This is obviously undesirable, and good resistance to autoxidation is a primary quality factor for these oils.

In usual practice, spray oil is applied at a level of 15% of the weight of the baked cracker. The recent trend toward reducing fat in the diet has led to work aimed at reducing this number. Some studies have shown that 5% spray oil gives a cracker with acceptable eating quality, with a significant reduction in both caloric density and the percentage of calories coming from fat.

PIE CRUST, BISCUITS

Two factors are important in making flaky, tender pie crusts and biscuits. First, the fat is divided into lumps in the dry flour, not uniformly incorporated. Second, after the addition of water, further working of the dough is minimized to prevent excessive gluten development. The usual cookbook description of the first step is "cut the fat into the flour until it is pea-sized." In commercial production, this means that the mixing must be done with a cutting action, and it must not be overdone. The mixing time depends in part on the shortening characteristics. The fat should have a moderate SFI curve, so that the mixing element can subdivide it easily. Also, the fat should be rather brittle, so that it shatters into small pieces. Often it is recommended that the fat be kept refrigerated and added to the mixer cold. If the flour and fat temperatures are too warm at the mixer, the fat will spread ("grease out") on the flour particles, and the resulting crust will lose flakiness.

Biscuit dough is worked more than pie dough after the liquid is added, but this mixing should be no more than enough to allow the dough to be gently sheeted and biscuits to be cut. Again, it is important to keep the dough cool so that the fat remains in particles rather than greasing out on the flour. The fat used nowadays is vegetable shortening with an SFI appropriate for the product; however, the traditional fat for both pie crust and dough was lard. Lard has an SFI curve about like the pie-crust shortening in Table 5-1, but it is a β-tending fat. Thus, it is brittle, rather than smoothly plastic, and has the desired characteristics for these baked products.

FILLER FATS AND ICINGS

Fillings for sandwich cookies or sugar wafers and creme icings for cakes consist primarily of fat and sugar, with flavor and color added as desired. The consistency of the filling is determined to a large extent by the SFI profile of the fat. This profile must meet three requirements:

- The blend must have a soft consistency, so that it can be extruded onto the basecake or wafer or spread onto the cake.
- The filling or icing must be firm at room temperature and below, so that the *basecakes* or wafers don't slide when the cookie or cake piece is eaten.
- The fat must melt almost completely at mouth temperature, so that it does not have a waxy mouthfeel.

Basecake—The baked product or surface to which a filling or icing is applied.

Sandwich cookies. Fat usually represents about 30–40% of the filling, and the solid fraction of filler fat gives body to the filling. The plastic range of filler fat is narrow. The mixed filling is somewhat warm (around 30°C or 86°F) for easy extrusion. When the sandwich cookie is cooled to room temperature, the cream sets up to the firm consistency needed to prevent sliding of the basecakes in the package. To achieve these goals, the SFI profile of a filler fat is rather steep, higher than that for all-purpose shortening at low temperatures and lower at high temperatures. The SFI specifications given in Table 5-1 for sandwich cookie and wafer filler fats are only typical examples. Bakeries will often adjust their specifications upward or downward to compensate for differences in temperature between seasons or geographical locations and to fit their equipment and processing conditions.

Sandwich Cookie Filling Fat Specifications

AOM stability: 100 hr
Oil stability index: 40 hr
Wiley mp: 39 ± 1°C

Solid fat profiles:

°C	SFC	°F	SFI
10	53 ± 4	50	38 ± 3
15	39 ± 3	70	24 ± 2
20	28 ± 3	80	18 ± 1
25	21 ± 2	92	9 ± 1
30	11 ± 1	104	<2
35	5 ± 1		
40	<1		

Sugar wafers. Sugar wafer filling is similar to sandwich cookie filling, except that finely ground scrap from the wafer trimming operation is generally added. In the production of filled wafers, it is crucial that the fat crystal structure be β'. If the shortening has started to convert to β crystals, the resulting oiliness causes the wafer sheets to slide during transport and cutting. Also, β crystals set up more slowly than β' crystals, causing delays between the extruding and cutting operations.

Wafer Filler Fat Specifications

AOM stability: 100 hr
Oil stability index: 40 hr
Wiley mp: 35 ± 1°C

Solid fat profiles:

°C	SFC	°F	SFI
10	76 ± 6	50	55 ± 4
15	60 ± 5	70	39 ± 3
20	45 ± 5	80	29 ± 3
25	32 ± 4	92	4 ± 1
30	17 ± 2	104	<1
35	1 ± 1		
40	<1		

Cake icings, fillings. Cake icings and fillings are similar in formulation to cookie fillings, but they are typically more highly aerated. Cookie fillings may have a specific density in the range of 0.8–1.1 g/ml, whereas creme fillings for cakes are often whipped to a specific density of 0.5. This aeration is promoted by emulsifiers in the shortening. Monoglycerides are commonly used, but, surprisingly, a high level of monoglyceride decreases the stability of the aerated filling. The optimum level is around 2.5% in the shortening. Addition of other emulsifiers, for example 0.9%

Polysorbate 60, results in an icing that is even more stable.

Sweet goods icings. Sweet goods icings are basically a suspension of very fine sugar crystals in a saturated sugar syrup. Gums are used to stabilize the system and inhibit growth of sugar crystals (which would give a gritty mouthfeel). High-melting fat hard flakes (with an iodine value [IV] of 5, melting point around 57°C, 135°F) amounting to 1–2% of the weight of sugar are often included during makeup of the icing. The fat contributes flexibility to the cooled icing and keeps it from flaking off the sweet goods.

Pan-Release Agents

Fat or oil is sometimes applied to the surface of bakery pans to facilitate the removal of the baked product from the pan. Two types are used.

Food grade mineral oil blended with partially hydrogenated soy oil (30:70) is used to lightly spray the surfaces of bread and bun pans that are in contact with dough. The maximum allowable amount of residual mineral oil in the product is 0.15%. The spray applicator must be calibrated to make sure that this level is not exceeded. The IV of the soy oil should be no higher than 40. This minimizes the amount of polymerization due to oxidation of unsaturated fatty acids and hence the buildup of "*varnish*" on the pan.

Varnish—An insoluble residue that results from the polymerization and buildup of fat on a hot surface.

In the second type, a solid material such as soy flour is added to vegetable oil material to increase its viscosity. This mixture is used to coat the inner surfaces of layer cake pans. A typical formulation contains 30% soy flour, 30% all-purpose shortening, and 40% RBD oil. This mixture, which is semiliquid at room temperature, is applied to the cake pan with a brush. (Spraying is not successful because the solids rapidly clog the nozzle.) The disadvantage of this release agent is that the oil rapidly polymerizes in the oven, and after fewer than 100 uses a rather heavy deposit of varnish builds up on the pan. The advantage is that a retail baker can readily make it from ingredients available in the bakery and doesn't have to purchase it in large amounts from a supplier.

In commercial bakeries, the pans used have been glazed by a company specializing in providing this surface. The glaze is a thin layer of glassy material that provides for ready release of the baked product. If the pans are properly handled, the glazing should be effective for 500–1,000 cycles. This approach is more cost-efficient in the long run.

Troubleshooting

The key to troubleshooting shortening-connected problems with bakery products is understanding the proper function of shortening in promoting the desired product characteristics, then determining to what extent the problem reflects incorrect shortening properties. The trouble spot may be anywhere in the entire process (formulation, mixing, processing, baking, or storage), so a certain amount of detective work is required to pinpoint changes that might help solve the problem. The troubleshooting table starts with observable symptoms. Table 5-1, showing typical bakery shortening properties, will help in determining some of the causes of those symptoms.

BREAD		
Symptom	**Causes**	**Changes to Make**
Sidewall weakness ("keyhole" cross section)	Low crumb structural strength and elasticity	Increase hard fat in shortening to at least 0.3% (flour basis). Add dough strengthener (SSL, DATEM) up to legal limit of 0.5% (flour basis). Increase amount of flour in the liquid preferment.
Open grain	Inadequate gluten strength and development	Check to make sure mixing time is optimum for the flour and formulation. Add dough strengthener (SSL, DATEM) up to legal limit of 0.5% (flour basis).
Low volume (sometimes noted when substituting oil for plastic shortening)	Inadequate gluten strength	Add dough strengthener (SSL, DATEM) up to legal limit of 0.5% (flour basis)
	Poor oven-spring	Increase shortening in formula to at least 3% (flour basis).
LAMINATED DOUGHS		
Symptom	**Causes**	**Changes to Make**
Dough tearing during roll-in	Shortening much harder than the dough	Use shortening with a lower SFI profile.
	Shortening is brittle	Make sure shortening is in proper β' phase (not gone to β due to warm storage temperatures, or improperly tempered by manufacturer).
	Dough too soft	Decrease dough absorption. Make sure dough is adequately retarded before rolling in.
Loss of flakiness in final product	Shortening completely melting in proofbox	Use a shortening with a melting point higher than proofing temperatures.
	Shortening is too soft during roll-in	Use shortening with a higher SFI profile. Return dough-fat "book" to retarder between sheeting steps.
	Insufficient number of layers in final dough	Increase number of sheeting steps.

YEAST-RAISED DOUGHNUTS

Symptom	Causes	Changes to Make
Poor surface appearance	Doughnut too shiny	Use frying fat with a higher SFI at room temperature.
	Doughnut too dull	Use frying fat with a lower SFI at room temperature.
	Doughnut too dark	Fryer temperature is too high.
	Doughnut too light	Fryer temperature is too low.
	Doughnut has blistered surface	Fryer temperature is too high.
		Doughnut is overproofed before frying.
Shelf life is too short	Inadequate fat absorption	Lower temperature of frying fat, and fry longer. Be sure frying fat is adequately tempered (FFA level 0.25–0.75%). Add 0.5% monoglyceride to base dough.

CAKES

Symptom	Causes	Changes to Make
Tunnels in the cake	Inadequate emulsification	Use a shortening with a good level (3.5–5%) monoglyceride. Add emulsifier with a high HLB (e.g., 0.25% Polysorbate 60) to shortening.
	Batter too warm	Cool batter during mixing.
	Batter viscosity too low	Decrease amount of water in batter formula.
Low volume, with a dip in the center	Inadequate emulsification	Use a shortening with a good level (3.5–5%) of monoglyceride.
	Air bubbles in batter are too large (not adequately dispersed during mixing)	Increase first-stage creaming (three-stage production). Increase hold time in mixing head (continuous cake production).
	Poor foam stability (in oil cakes)	Increase level of α-tending emulsifier (PGME, AcMG) to 12–14% of oil shortening. Add 1–3% soluble protein source (egg white, milk protein).
	Cake flour gelatinizing too late during baking	Use cake flour with a lower flour pH.
Excessive peaking, top cracking	Insufficient shortening	Increase shortening level in base formula.
	Cake batter viscosity too high	Increase amount of water in batter formula.
	Cake flour gelatinizing too early during baking	Use cake flour with a higher flour pH.
Texture problems	Cake is tough	Increase shortening level in base formula.
	Cake is too tender	Decrease shortening level in formula.

CAKE DOUGHNUTS

Symptom	Causes	Changes to Make
Excessive fat absorption	Cold batter	Use warmer water during batter make-up.
	Batter viscosity too high	Increase amount of water in batter formula.
	Under mixed	Increase mixing time to fully hydrate ingredients.
	Fryer temperature too low	Raise fryer temperature to recommended level. Make sure heater can keep up with rate of batter deposition.
	Fat breakdown (high FFA)	If fryer not in continuous use, lower fryer temperature between fries. If continuous, match doughnut production with fryer size to get complete fat turnover in 8 hr.
	Excess emulsification	Reduce emulsifier amount in doughnut formula.
Low fat absorption	Hot batter	Use cold water (or ice) to reduce batter temperature.
	Batter viscosity too low	Reduce water in batter formula.
	Over mixed	Decrease mixing time.
	Too much floor time before depositing	Balance batch size and mixing schedule with rate of production in fryer.
	Fryer temperature too high	Lower fryer temperature to recommended level.
	New (untempered) fat	When refilling fryer (after cleaning) use old (not degraded) fat equal to one-fourth of fryer capacity. If old fat is not available, add 1½ oz free fatty acids per 100 lb new fat (0.1% FFA).
Sugar won't adhere	Frying fat SFI too high	Use frying fat with lower SFI at room temperature.
	Doughnut sugar fat SFI too high	Use a fat in the sugar with lower SFI profile.
	Doughnuts too cool when coating	Shorten cooling time before doughnuts enter sugaring tumbler. Warm cooler air, with infrared heaters.
Sugar disappears during shelf life	Frying fat SFI too low	Use frying fat with higher SFI at room temperature.
	Doughnuts too warm when sugared	Lengthen cooling time before sugaring. Cool air in cooler.
	Excess moisture in package	Cool doughnuts before packaging. Do not store in warm warehouse. Use packaging material that is more water permeable.
Poor surface appearance	Doughnut too shiny	Use frying fat with a higher SFI at room temperature.
	Doughnut too dull	Use frying fat with a lower SFI at room temperature.
	Doughnut too dark	Fryer temperature too high.
	Doughnut too light	Fryer temperature too low.
	Doughnut has a blistered surface	Fryer temperature too high.

COOKIES

Symptom	Causes	Changes to Make
Rotary molded cookies don't release from die	Shortening SFI too high	Use shortening with a lower SFI profile. Substitute oil for a small portion of shortening.
	Dough too cold	Adjust water temperature to raise dough temperature.
Wire cut cookies extrude unevenly	Inadequate lubrication Insufficient shortening	Increase oil in dough formula.
	Shortening is too hard	Use shortening with a lower SFI profile. Substitute oil for a small portion of shortening.

SNACK CRACKERS

Symptom	Causes	Changes to Make
Rancid odor, taste	Spray oil is oxidizing	Use a spray oil with AOM of 100 hr minimum. Specify addition of antioxidants (BHA, BHT, TBHQ) to spray oil.
	Metal ion contamination of spray oil	Specify addition of chelator (citric acid) to spray oil.
	Improper bulk storage of oil	Fill headspace of storage tanks with N_2 gas.
Surface appearance Too shiny	Spray oil melting point too low	Use spray oil with complete mp of about 30°C (86°F).
Too dull	Spray oil SFI too high at room temperature	Use spray oil with SFI about 5 at 25°C (77°F).

PIE CRUST, BISCUITS

Symptom	Causes	Changes to Make
Tough, nonflaky	Too little fat	Increase amount of shortening in formula.
	Dough overworked	Decrease mixing time; mix just long enough to incorporate all ingredients. Use minimum amount of sheeting to achieve desired dough thickness. Minimize amount of rework material back to extruder.
	Flour too strong	Use flour with a lower protein level, or softer type of flour.
Fat disappears during processing	Shortening SFI too low	Use shortening with higher SFI profile.
	Shortening too warm when added to mixer	Cool (temper) shortening before adding to mixer.
	Dough too warm	Use cool water (or ice) for lower dough temperature.

SANDWICH COOKIES

Symptom	Causes	Changes to Make
Difficult to deposit evenly	Filler fat SFI too high	Use filler fat with a lower SFI profile.
	Filling too cool	Warm filler fat and/or sugar before adding to mixer.
Leaves waxy taste	Filler fat melting point too high	Use filler fat with complete mp no higher than 39°C (102°F).

SUGAR WAFERS

Symptom	Causes	Changes to Make
Difficult to deposit evenly	Filler fat not plastic; "gone beta"	Store fat at temperatures below 33°C (92°F).
	Filling too cool	Warm filler fat and/or sugar before adding to the mixer.
Wafers slide before cutting	Filler fat not stabilized in β′ crystals	Make sure supplier is using β′-stable source fats. Make sure supplier is properly tempering shortening after plasticization.
	Wafers too cool at depositor	Shorten time in cooler, or raise cooler temperature.
	Insufficient cooling before cutting	Lengthen time in cooler, or lower cooler temperature.
Leaves waxy taste	Filler fat melting point too high	Use filler fat with complete mp no higher than 36°C (97°F).

CREME ICINGS

Symptom	Causes	Changes to Make
Specific gravity too high	Inadequate emulsification	Use shortening with 2–3% monoglyceride. Use shortening with some high HLB emulsifier, such as 0.9% Polysorbate 60.
	Fat too soft to entrap air	Use shortening with higher SFI profile.
	Inadequate creaming	Increase first stage mixing time of fat plus sugar.
Specific gravity too low	Excessive whipping	Decrease mixing time after all ingredients are added.
Loses aeration during storage	Excess monoglyceride	Use shortening with maximum 3% monoglyceride.
	Inadequate emulsification	Include high HLB emulsifier such as Polysorbate 60 in shortening, up to 0.9% level.
	Filler fat not stabilized in crystals	Make sure supplier is using stable source fats. Make sure supplier is properly tempering shortening after plasticization.

Frying Fats

In frying, heat energy is transferred from the heat source to the food being cooked via a fat. Most industrial processes involve deep-frying, i.e., the heat transfer medium is bulk fat, and the food is partially or totally submerged. In addition to changes in the food brought about by heat, numerous chemical reactions change the physical nature of the fat itself and affect its ability to make a finished product with the desired characteristics.

Heat Transfer by Fat

MASS TRANSFER

During frying, water leaves the product being fried and is transferred to the atmosphere above the fryer as steam. In addition, a certain amount of the frying fat enters the product. These two apparently simple processes (evaporation and absorption) are key determinants of the frying process and largely govern whether or not the final product is of good quality or is deficient in one or more characteristics.

Water evaporation. Typically, frying is done at 180–190°C (356–374°F). When the raw product is placed in the fryer, the temperature of the surface region quickly increases, causing water in this layer to evaporate. However, it does not simply "disappear" from the *interfacial region*. The product components (proteins, carbohydrates) are more hydrophilic than the frying fat, and water vapor does not readily migrate from this hydrophilic surface to the more *hydrophobic* bulk fat. Thus, at least in the initial stages, there is a thin layer of steam between the fat and the product (Fig. 6-1A). This steam serves as a partial insulator, because its heat conductivity is much lower than that of fat or of the still-liquid water in the food.

The steam layer is unstable, and portions of it enter the fat, then make their way to the surface and evaporate. This allows fat to again contact the surface of the food and transfer more heat energy, which allows the cooking process to continue (Fig. 6-1B). The disruption of the steam layer is aided by the presence of surface-active materials, from either the food (e.g., lecithin in a doughnut batter) or the frying fat (e.g., fatty acids or oxidized fat). When the food is essentially solid, for example, a potato piece (french fries), the steam can escape only to the fat phase. In cases where the food contains internal voids, such as a leavened doughnut, steam on the side of the void closer to the heat

In This Chapter:

Heat Transfer by Fat
  Mass Transfer
  Temperature Gradients
  Optimum Fryer Operation

Degradative Reactions
During Frying
  Hydrolysis
  Oxidation
  Polymerization
  Smoking, Ignition

Selection of Frying Fat
  Stability
  Suitability

Troubleshooting

Interfacial region—The area where two dissimilar materials are in contact with each other.

Hydrophobic—Water hating or nonpolar.

Typical frying temperature: 180–190°C (356–374°F).

Micelles—Structures in which similar or dissimilar molecules are arranged in an orderly manner.

> Surface-active free fatty acids help heat transfer to the product but also promote hydrolysis and thus degradation of the fat.

source (i.e., fat) can recondense on the other side of the void, thus effectively transferring heat energy to the interior of the doughnut.

Fat-soluble emulsifiers aid the transfer of water from product to fat by promoting formation of water *micelles* in the fat. (These are small water-in-oil emulsions.) In hot fat, they are extremely tiny, but when frying fat is cooled after use they often appear as a haze. The significance of micellization is that it helps water penetrate the hot fat and promotes fat hydrolysis. The surface-active free fatty acids, then, are a two-edged sword; they promote heat transfer to the product, but they also promote fat degradation.

Fat absorption. As water migrates out of the food, it leaves voids on the surface of the solid food components along which a thin layer of fat migrates (Figure 6-1B). This process enhances heat transfer as discussed above. The fat remains in the product interstices after removal from the fryer and forms a significant part of the finished product. Table 6-1 gives the content of certain foods before and after frying.

The characteristics of the frying fat are important aspects of product quality. In doughnuts and fried snacks, the absorbed fat tenderizes the product. In all cases, the flavor of the product is strongly influenced by the flavor of the fat. The solid fat index (SFI) profile of the fat also affects eating quality. If the solids content of the fat is rather high at body temperature, it will leave a waxy mouthfeel. On the other hand, if the SFI is low at room temperature, the product will look greasy or oily. If the fat lacks adequate stability, or if it has been abused by improper management of the frying equipment (see the section on fryer operation below), product crust color may suffer and off-flavors and odors may be introduced.

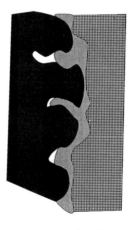

A. Fresh Fat

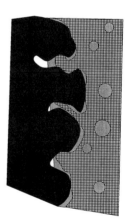

B. Optimum Fat

■ Food solids □ Water ▨ Steam ▦ Frying fat

Fig. 6-1. Distribution of materials at the food/fat interface during deep frying. A, Orientation of steam, food surface, and fat in the absence of surface-active materials. B, How emulsifiers help the steam layer leave the food and aid in fat absorption.

For a given product, there is an optimum amount of fat absorption, determined as the "best tasting product." Manufacturers of fried foods establish a desired value (or range) for absorbed fat (part of the finished product specifications), using some combination of tradition, recommendations, and sensory panel testing. The three main determinants of degree of fat absorption are:

- Frying temperature
- Concentration of surface-active agents
- Water binders in the food.

Temperature. If the fat temperature in the fryer is too low, frying time is often extended to achieve the desired depth of crust coloration. However, this also allows more time for fat to penetrate into the interior of the product. Also, the pressure of the steam being generated is less than optimum, which also makes it possible for more fat to penetrate. We have all undoubtedly had the unhappy experience of being served pale, greasy french fries. This situation invariably means that the cook is not keeping good watch on the fryer thermostat (or that the fryer is being overloaded).

Surface-active agents. Emulsifiers promote breakdown of the steam barrier, and this allows the oil to wet the solid food surfaces. Both of these effects increase the absorption of fat. In some cases (e.g., cake doughnuts), a small amount of a wetting agent such as lecithin is added to the batter to help achieve the desired level of fat pickup. Also, free fatty acids generated by fat hydrolysis and oxidized fat components (see below) are surface active and promote fat absorption.

TABLE 6-1. Changes in Fat Content upon Deep-Frying

Product	Percent Fat	
	Raw	Fried
Chicken (no skin)	3.9	9.9
Ocean perch	1.2	13.1
Potato chips	0.1	39.8
French fries	0.1	13.2
Cake doughnuts	5.2	21.9

Water binders. Soluble proteins, gelatinized starch, gums and other fibers, and similar water-binding ingredients decrease fat absorption by keeping more water in the system. This property is used in formulating batters and breadings for coating foods that are to be deep-fried. Studies have shown, for example, that the inclusion of 3% cellulose fiber in a doughnut formulation markedly decreases fat absorption.

TEMPERATURE GRADIENTS

Several temperature gradients exist within a deep-frying operation, all of which affect the final product. The obvious one, of course, is from the fat to the center of the food being cooked. This is a large gradient, ranging from 25°C (initially) at the core of room-temperature food to 185°C in the bulk fat. As heat penetrates the food, it cooks (with all the changes that accompany that process). However, the core temperature never exceeds 100°C and more usually is 80–95°C, depending upon the product (for example, chicken temperature is at the lower end of that range and cake doughnut temperature is at the upper end of it).

A second gradient is in the fat near the region where product is introduced. If the product is cold and there is a large amount of it relative to the amount of fat in the fryer, the local temperature may drop by more than 30 degrees C (54 degrees F). This is an undesirable situation because some time is required before heat energy can be conducted from the heat source into that area. During this interval, cooking proceeds at suboptimum temperatures. A better situation exists in continuous fryers, such as wholesale doughnut fryers or corn chip fryers, where the raw product is introduced in continuous small increments at one end of the machine and is conveyed the length of the fryer (Fig. 6-2). With this arrangement, the temperature at the introduction point may be no more than 10 degrees C (18 degrees F) lower than in the main part of the fat.

A third gradient, often overlooked, is between the heat source and the bulk fat. The heat source is usually a gas flame burning in a tube. The heat is transferred through the metal wall of the tube to the fat. Various configurations (fins, serpentine tubes, etc.) are used to improve this transfer. The transfer of heat is hindered if the surface of the heat exchanger is coated with food particles and polymerized fat. However, a rather large temperature gradient exists between the fat immediately in contact with the heat exchanger surface and the fat in contact with the food product. This gradient is lessened if some sort of undersurface forced circulation system is used. For example, fat from near the heat exchanger may be pumped through a filter element, then reintroduced into the fryer well below the surface (to prevent any aeration).

OPTIMUM FRYER OPERATION

Fat condition stages. The progressive status of fat in a fryer is often divided into six stages, as shown in Table 6-2. Each stage has certain

Fig. 6-2. A continuous doughnut fryer, in which doughnuts are introduced at one end of the frying compartment, then are conveyed through the hot fat and exit at the other end. (Courtesy DCA Food Industries, Jessup, MD)

implications for the frying process and product quality. Frying in new fat is very difficult. The lack of surface-active materials hinders steam escape, fat absorption, and heat transfer. Break-in fat works somewhat better but still gives inadequate fat absorption. Fresh fat gives good quality; fat at this stage is optimum for cake doughnuts, for example. Optimum fat gives the best performance in most restaurant frying of french fries, meats, vegetables, etc. Degraded fat is "over the hill" for good frying, usually giving excessive crust coloration and fat absorption. In many countries, a fat with a polar lipid content greater than 27% is now illegal, and fat that has reached this stage must be discarded. Runaway fat, of course, has been subjected to excessive degradation; it should never be seen in properly managed fryers.

Fat turnover. As fat is removed from the fryer by absorption, it is replaced with more fat. The food removes polar lipids (oxidized fat, polymers, free fatty acids) along with triglycerides, and these are all replaced with nearly pure triglycerides. The fryer should be managed so that this replacement serves to keep the oil in the desired stage for good frying. The size of the fryer (fat capacity) should be matched to the rate of food throughput (which affects fat absorption). For commercial doughnut frying, for example, this match should give one complete fat turnover during one 8-hr shift. Other products having different fat absorption and throughput rates would have different turnover rates in a fryer of the same capacity. The time for one complete turnover should be no more than 24 hr.

In many operations (e.g., restaurants and fast-food stores), frying is intermittent, not continuous. When the fryer will not be in operation for a period of time, the heat should be turned down to about 155°C (310°F). Lowering the fat temperature by 30 degrees C decreases the rate of degradation eightfold. This extension of useful fat life more than compensates for the delay in reheating the fat before the next frying operation.

Filtration. During frying, solid particles appear in the fat. These may be pieces that have broken off from the surface of the food or soluble protein or starch carried into the fat (along with evaporating water) and then carbonized. These particles collect on heat exchanger surfaces, interfering with heat transfer. They also contribute to continuing degradation reactions. They should be removed by *passive filtration*,

> To extend the life of the fat, the fryer should be turned down to about 155°C (310°F) when not in use for an extended time.

Passive filtration—Running material through a filter using gravity.

TABLE 6-2. Stages in Frying Fat Condition

Oil Stage	Description	Triglycerides (%)	Polar fat (%)	Polymers (%)	Free Fatty Acids (%)
New fat	No surface-active materials	>98	<2	0.5	0.03
Break-in	Some surface-active materials	90	10	2	0.5
Fresh	Best for cake doughnuts	85	15	5	1
Optimum	Best for meats, french fries, vegetables, etc.	80	20	12	3
Degraded	Much surface-active material; is easily absorbed	75	25	17	5
Runaway	Very degraded	65	35	25	8

using a steel mesh or filter paper, sometimes with the addition of diatomaceous earth (filter aid). This type of filtration is applied to fat during the break-in and fresh stages.

Active filtration is applied to fat during the optimum and degraded stages. Adsorbent materials that remove soluble compounds (polar lipids, polymerized materials) extend the optimum lifetime of the fat. Examples of such materials are α-cellulose, silica gel, alumina, magnesium and calcium silicates, and activated carbon. Circulation of hot fat through the adsorbent bed should be continuous for maximum effectiveness.

Active filtration—Using pressure to force material through a filter.

Degradative Reactions During Frying

Fat, which is composed of triglycerides and contains unsaturated fatty acids, is subject to several chemical reactions that change its characteristics. The rate of these reactions doubles for each 10 degrees C (18 degrees F) of increase in temperature, so lowering the temperature of fat when the fryer is not in use increases its useful lifetime.

HYDROLYSIS

Water, introduced into the fat from the food, causes hydrolysis of the ester bonds in the fat, producing diglycerides, monoglycerides, and free fatty acids. The latter two materials are emulsifiers, which, as noted above, are promoters of fat hydrolysis. Micellization of water keeps it in the fat phase for a longer time, which also increases hydrolysis. In addition, hydrolysis is catalyzed by alkaline materials. For this reason, the fryer must be thoroughly rinsed with acid (e.g., vinegar) and water after being cleaned with alkaline cleaning aids.

OXIDATION

Oxidation at the double bonds of unsaturated fatty acids leads to the formation of a variety of polar compounds: aldehydes, ketones, acids, peroxides, etc. These materials, which are all surface active, change the frying performance. The aldehydes and ketones react with proteins in the food and increase crust color formation. Oxygen from the air is the reactive species; it is introduced into the fat in several ways:

- Splashing, as product or fresh fat is added to the fryer
- Turbulence, when foam bubbles break
- Air introduced through leaks in the continuous filtration apparatus
- Diffusion at the surface of the fat.

Metal ions (e.g., iron, copper, and manganese) catalyze oxidation reactions. Copper is particularly effective in this regard and should never be used for frying equipment. In particular, breaks in screens, chains, and other parts must never be repaired using any type of brazing.

POLYMERIZATION

Unsaturated fatty acids can be chemically joined (polymerized) by various reactions. Oxidation products (for example, epoxides) can readily react to form dimers. Polyunsaturated chains can undergo direct polymerization. The practical result is that the fat contains some percentage of polymerized triglycerides. As this process progresses, the fat can actually form a varnish that deposits on the metal surfaces of the fryer. The polymeric materials are foam stabilizers; as their concentration increases, the fat foams more readily when food is fried. As with oxidation, polymerization is catalyzed by metal ions.

SMOKING, IGNITION

As the free fatty acid concentration in frying fat increases, the temperature of the smoke, flash, and fire points (see Chapter 2) decrease.

Notice back in Figure 2-4 that the particular sample of fresh oil studied originally had a smoke point of 212°C (414°F) but that 1% free fatty acid lowered the smoke point to 152°C (314°F). The flash point was about 138 degrees C higher and the ignition point 56 degrees C higher yet (250 and 100 degrees F). This particular work used the direct addition of free fatty acids to make these measurements. In an actual fryer, many of the oxidation products also lower these danger points, and fryer fires are by no means unknown. The best way to avoid them is to prevent degradation of fat.

Selection of Frying Fat

STABILITY

The chief characteristic required of frying fat is good stability with respect to oxidation and polymerization. This means that the content of unsaturated fatty acids, particularly polyunsaturated fatty acids, must be minimized, consistent with the SFI requirements for the fat in the finished product. For light-duty frying (home or occasional restaurant), where the fat is used once and then discarded, a refined, bleached, deodorized (RBD) vegetable oil will suffice. For heavy-duty, continuous frying, the oil must at least be lightly hydrogenated to remove linolenic acid; more often it is partially hydrogenated to de-

---

Specifications for Frying Oils

All Oils

- Peroxide value: 1 meq/kg maximum
- Free fatty acid (as oleic acid): 0.05% maximum
- Phosphorus content: 1 ppm maximum
- Color (Lovibond): 1.5 R, 15 Y maximum
- Flavor: Bland
- Odor: Neutral when warmed

Light Duty Frying Oil

- AOM stability: 25 hr
- Oil stability index: 10 hr
- Smoke point: 225°C
- Solid fat profiles:

°C	SFC	°F	SFI
10	<5	50	<5
15	<2	70	<1.5
20	<1	80	0
25	0		

High-Stability Frying Oil

- AOM stability: 200 hr
- Oil stability index: 80 hr
- Smoke point: 235°C
- Wiley mp: 38 ± 1°C
- Solid fat profiles:

°C	SFC	°F	SFI
10	65 ± 4	50	47 ± 3
15	51 ± 4	70	32 ± 3
20	37 ± 3	80	25 ± 2
25	24 ± 3	92	12 ± 1
30	16 ± 2	104	<2
35	7 ± 1		
40	<1		

crease the linoleic acid. Heavy-duty frying oils (melting point less than 25°C) are available. These are made by partial hydrogenation of vegetable oil, followed by fractionation to remove high-melting solid fractions. Partially hydrogenated vegetable oils with melting points from 35 to 45°C (95–113°F) are used, for example, in doughnut and snack frying operations. Frying fats and oils with a wide variety of melting points, SFI profiles, and AOM specifications are available from commercial suppliers.

Additives increase fat stability and frying performance. Citric acid is commonly added to shortenings to chelate deleterious metal ions (iron, copper, manganese), to prevent their promotion of autoxidation. Antioxidants (BHA, BHT, TBHQ) are added to delay the onset of rapid peroxide formation. Finally, dimethyl silicone may be used to inhibit foam formation.

SUITABILITY

While fat stability is important to the operation of the fryer, properties of the fat itself influence the consumer appeal of the finished product and must be taken into account.

SFI profile and finished product. The solid fat content that absorbed fat has at the temperature of consumption influences its quality. For example, some potato chip makers believe their chips are more appealing if they have a dry surface appearance (SFI >10 at room temperature), while other manufacturers prefer a shiny appearance (melting point at or below room temperature). Similar differences of opinion are held with respect to many other breaded and battered fried foods.

On the other hand, for sugaring doughnuts, the absorbed fat must have the proper SFI at the temperature of application. If it is too hard the sugar won't adhere, while if it is too soft the oil will penetrate the sugar layers, giving a poor appearance.

Flavor. A second factor to consider is fat flavor. Most of these flavor notes arise from the interaction of fat with proteins and carbohydrates during frying. However, individual oil sources can also contribute characteristic flavors after some use. Thus, soybean and cottonseed oils are characterized by a metallic note; corn oil gives a mild corn note; peanut oil has a nutty note; canola and sunflower oils are very mild; and palm oil gives a sweet pumpkin flavor note. These are generally quite delicate compared to the other flavors in fried food, but if a slight "off-flavor" is detected in a product fried in RBD or lightly hydrogenated oil, it may be worthwhile to try some other oil source.

Troubleshooting

FAT ABSORPTION		
Symptom	**Causes**	**Changes to Make**
Excessive fat pickup	Low fryer temperature	Make sure thermostat is properly calibrated. Decrease food load; use smaller batches or slower addition. Increase capacity of heating element Clean heat element tubes to prevent insulation by food particles or polymerized fat.
	High free fatty acid content in oil	Increase rate of production to increase fat turnover.
	Improper food formulation	Decrease emulsifier level in formula. Add protein source to formula. Use water-binding fiber sources in food or coating.
	Degraded oil	Replace oil more often. Cool fryer between uses. Include an active in-line filter in the system.
Inadequate fat pickup	High fryer temperature	Make sure thermostat is properly calibrated.
	Low free fatty acid content in oil	Lower production rate. If necessary, add 0.1–0.3% free fatty acids.
	Low emulsifier content	Increase emulsifier level in formula.
	Fresh, untempered oil	When refilling fryer after cleaning, use at least 25% tempered oil.

PRODUCT COLOR		
Symptom	**Causes**	**Changes to Make**
Product is too light	Low fryer temperature	Make sure thermostat is properly calibrated. Decrease food load; use smaller batches or slower addition. Increase capacity of heating element. Clean heat element tubes to prevent insulation by food particles or polymerized fat.
	Excess moisture in food	Drain thoroughly before adding to fryer.
	Food too cold when added to fryer	Thaw frozen food before adding to fryer.
	Cooking time too short	Adjust cooking cycle, production timing.
Product is too dark	High fryer temperature	Make sure thermostat is properly calibrated.
	Degraded oil	Replace oil more often. Cool fryer between uses. Include an active in-line filter in the system.

FAT PROBLEMS

Symptom	Causes	Changes to Make
Foaming	Overheating	Make sure thermostat is properly calibrated.
	Inadequate turnover	Increase ratio of food load to fryer capacity.
	Food particles present	Filter fat more often, or install continuous filter. Clean fryer more often; be sure proper methods are used.
	Formation of polymerized fat	Include chelators (citric acid) in frying fat. Prevent splashing when food is loaded or removed. Lower fat temperature when fryer is not in use. Install active-filter continuous filtration system.
Smoking	Fryer is too hot	Make sure thermostat is properly calibrated. Set thermostat to lower temperature.
	Fatty acid content too high	Remove excess water (or ice) from food before adding. Increase rate of fat turnover. Lower fat temperature when fryer is not in use. Ensure adequate rinsing after cleaning; use acidified water for at least one rinse.
	Food particles present	Filter fat more often, or install continuous filter. Clean fryer more often; be sure proper methods are used.

Chocolate and Confectionery Coatings

Function of Fats and Oils in Enrobing Compounds

FLAVOR, TEXTURE

Confectionery coatings include all of those fat-based formulations that are melted and applied to some sort of solid food and then solidify to form the coating. The process, generally called *enrobing*, is applied to a wide variety of foods: nuts, nougats, flavored gels, cakes, doughnuts, cookies, sugar wafers, ice cream bars, fruit pieces; and almost anything that the manufacturer thinks will be enhanced by the addition of the flavor and texture of the coating.

The primary purpose of the coating is to be a flavor carrier, and the most popular flavor is chocolate. When other flavors are used, the carrier fat is either *cocoa butter* alone or a bland fat derived from another source, with flavors (e.g., vanilla or strawberry) added.

The coating fat has two primary textural requirements: it must be rather hard at room temperature so that it doesn't soften or melt during storage and handling, and it must have a melting point close to body temperature so that it doesn't leave a waxy mouthfeel. In these respects, cocoa butter is considered the model, although its melting point (32–35°C) is a little low for some purposes. When used to coat baked products, confectionery fat melts in the mouth and provides the desired lubrication and palate-clearing properties.

FAT CRYSTALLIZATION

Proper crystallization of the fat on the enrobed piece is important to finished product quality. The main requirement is that the surface be glossy, with the desired shade of brown. Chocolate coating will often develop a dull, dusty white appearance during storage, known as *"bloom."* This phenomenon comes from changes in the polymorphic crystal structure of the fat. It has many causes: improper tempering of the coating, the use of incompatible fats, improper conditioning of the centers before coating, poor temperature conditions during enrobing, large temperature fluctuations during storage, and others. The proper control of these factors to prevent blooming is discussed below (see the sections on usage and troubleshooting).

In This Chapter:

Function of Fats and Oils in Enrobing Compounds
   Flavor, Texture
   Fat Crystallization

Chocolate Coating
   Origin, Manufacture, Processing
   Chocolate Types
   Usage
   Extenders

Compound Coating
   Lauric Fats
   Other Fat Sources

Troubleshooting

Enrobing—The process of covering a base food material with a melted coating that hardens to form a solid surrounding layer.

Cocoa butter—The fat from cocoa beans used in chocolate. It has a sharp melting point just below body temperature.

Bloom—A dusty white appearance on the surface of chocolate caused by the formation of certain types of fat crystals.

TABLE 7-1. Polymorphism and Melting Points of Cocoa Butter

Phase	Form	Melting Point, °C
γ	I	17.3
α	II	23.3
Mixed (α + β′)	III	25.5
β′	IV	27.5
β	V	33.8
Segregated (β + β)[a]	VI	36.3

[a] β crystals with two different melting points.

In Chapter 4, fats were described as undergoing two crystal phase transitions, from α to β′ to β, during plasticization of melted fat. This sequence applies to nearly all edible fats, with one major exception, namely cocoa butter, which has at least four main crystal structures. Rapid cooling of fully melted cocoa butter produces a highly disordered crystal called the γ phase; upon subsequent slow heating, the transition proceeds through α, β′, and β crystals. Two other forms have been identified by X-ray crystallography. The first is now considered to be a mixture of α and β crystals, and the second occurs when the β phase segregates into lower- and higher-melting fractions. In addition to having Greek names, these forms are designated by the Roman numerals I–VI, which can be confusing. Table 7-1 summarizes the relationships and melting points for the various phases.

In chocolate, the preferred crystal phase is the stable β (V) form. If the chocolate is not tempered properly, or if other operating parameters (discussed below) are not correct, the cocoa butter crystallizes in one of the intermediate forms. Phase transformations occur during storage, and the surface appearance changes from smooth and glossy to dull. Eventually the whitish bloom appears. However, in coatings made with fats other than cocoa butter, the fat always crystallizes in the β phase. Because of the stability of this phase, such coatings are resistant to bloom formation.

Chocolate Coating

Chocolate is the consuming public's most-favored food flavor. Chocolate is the favorite flavor of ice cream; chocolate chip cookies are the number one seller; most candy bars are enrobed with a chocolate-flavored coating; and fine milk or dark chocolate is considered the ultimate in elegant confections. Because chocolate is a high-fat item, its successful use in food manufacture depends on proper manipulation of its fat properties.

ORIGIN, MANUFACTURE, PROCESSING

Cacao trees (*Theobroma cacao*) grow around the world in a belt between 20°N and 20°S latitude. Although these trees are native to the western hemisphere, 70% of the world's supply is now grown in West Africa. The three subspecies of *T. cacao*, influenced by genetic makeup and climatic variations, produce cocoa beans with differences in color and

Fig. 7-1. Cocoa beans on the tree. (Courtesy Bruce Stillings)

flavor, which are blended by cocoa processors to make the wide variety of available cocoas.

Cacao trees bear pods with a thick outer shell (Fig. 7-1), containing 20–40 beans (seeds) embedded in a thick mucilage. The ripe seed pods are cut from the trees, split open, and set on racks for three to five days, during which natural fermentation converts the mucilaginous gums into alcohol and vinegar. This drains away, leaving the moist beans, which are then sun-dried (Fig. 7-2) to a moisture content of 7–9%. The dry beans are bagged and shipped to the processor.

After destoning, the beans are roasted. A light roast gives a product with an acidic flavor, while a heavy roast yields a strong flavor. Most beans receive a medium roast. The shells are removed, leaving the *nibs*. These are ground in crushers, where frictional heat melts the fat. The liquid mass is poured into molds and allowed to solidify. The product is *chocolate liquor*.

Fat makes up over half the weight of chocolate liquor (Table 7-2). This fat is cocoa butter. By mechanical pressure, some of the fat is removed. The solid residue is *presscake*, which, when ground, yields cocoa powder. Removal of 75% of the fat gives presscake that contains 22–24% fat, which is used to make breakfast cocoa, to flavor dairy products, and in home baking. Removing 83% of the cocoa butter yields a presscake having 15-17% fat, which is ground to make a cocoa often used by retail bakers. The cocoa powder usually purchased for wholesale bakery applications contains 10–12% fat, which is obtained by removing about 89% of the cocoa butter from the chocolate liquor.

Cocoa flavor and color reside in the solid portion of chocolate liquor; the cocoa butter is nearly flavorless and colorless. It is also the more expensive portion. From 100 lb of chocolate liquor, the processor obtains approximately 52 lb of 10–12% fat presscake, selling for roughly $0.50/lb, and 48 lb of cocoa butter, worth about $2.50/lb. The unique

Nibs—The insides or kernels of cacao beans when the hulls are removed.

Chocolate liquor—The solid mass obtained when the ground, liquified cacao nibs are cooled.

Press cake—The solid residue remaining when cocoa butter is removed from chocolate liquor by pressing.

Fig. 7-2. Cocoa beans air-drying. (Courtesy Bruce Stillings)

Alkalized (dutched) cocoa— Cocoa products that are treated with an alkali agent to raise the pH and produce a darker color.

melting properties of cocoa butter are important in enrobing operations, but for most bakery applications other, less expensive, fats are preferable.

Alkalized (dutched) cocoa is obtained by treatment with food-grade alkali, usually potassium carbonate or sodium carbonate, at one of two stages. The nibs may be soaked in a solution of alkali before grinding. This is the traditional method and is considered to give a superior product. Alternatively, the presscake may be treated with alkali. This gives the same sort of color development and flavor modification and is less expensive, but the flavor is harsher. Addition of alkali to chocolate liquor causes some saponification of the cocoa butter, resulting in a soapy flavor, so the treatment is not made at this stage.

The dried beans have a pH of 5.2–5.6. The addition of alkali raises the pH to above 7. At higher pH, the phenolic materials in cocoa powder develop a reddish brown to black color; the higher the pH, the darker the color. Thus, ordinary (red-brown) dutched cocoa has a pH of 7.1–7.4, while "ebony" (black) cocoa has a pH above 8 and is used mostly for color. The dark crumb of devil's food cake results from the addition of enough soda to raise crumb pH to around 8.

The non-fat portion of chocolate liquor is fibrous in nature. The average particle size is reduced when the nibs are ground and may be reduced further, if necessary, during milling of the presscake. The particle size is called "fineness." It refers to the diameter of most of the particles (all but the largest ones), in units of ten-thousandths of an inch. Thus, in a cocoa with a fineness of 10, all but a few particles would be smaller than one-thousandth of an inch in diameter. Chocolate for confectionery and frozen dessert enrobing (in which a smooth mouthfeel is important) typically has a fineness of 5–7. For bakery use, where the cocoa powder or enrobing chocolate accompanies other components with coarser crumbs, a fineness of 10–12 is usual.

TABLE 7-2. Typical Composition of Some Cocoa Products

	Chocolate Liquor, Percent (Range)	Cocoa Powder, Percent	
		10-12% Fat, Natural	10-12% Fat, Dutched
Moisture	0.5–3.5	4.2	4.5
Protein	9–13	23.6	23.3
Starch	5–7	12.2	12.0
Fat (cocoa butter)	50–58	11.8	10.5
Ash (total)	2–4	6.0	8.4
Sugars	1–3	2.1	2.1
Cellulose	9–12	18.3	18.0
Polyhydroxyphenols	4–6	12.5	12.3
Pentosans	1.2–1.6	3.0	3.0
Organic acids	1.1–1.7	3.2	...
Theobromine, caffeine	1.2–1.7	2.8	2.8

CHOCOLATE TYPES

Chocolate liquor. Also called *"baker's chocolate"* or "unsweetened chocolate," chocolate liquor is used to give color and flavor to products such as chocolate cookies or chocolate ice cream. It must be melted before addition to the mixer but protected from high heat that would generate a burnt flavor and other off-flavors. Chocolate liquor melts at around 38°C (98°F); this is best done in a jacketed, continuously stirred container heated with hot water (no hotter than 60°C, 140°F). The melted liquor should be held at about 40°C (104°F) until it is added to the food in process.

Semisweet and *milk chocolate*. Chocolate liquor has a bitter flavor. Blending it with sugar, dry milk solids, lecithin, and vanilla and/or vanillin produces the appetizing chocolate ordinarily consumed. There is no standard designation for different sugar levels, so "semisweet" chocolate from different manufacturers may vary significantly in sweetness. In the absence of milk solids, the chocolate retains its native dark color, whereas milk components give a much lighter product. Some of the chocolate liquor used may be dutched, particularly for dark chocolate.

Solid chocolate is made by blending sugar, milk solids (if used), chocolate liquor, and some cocoa butter to create a semidry paste. This paste is ground on a five-roll refiner (to reduce particle size), then transferred to a slow mixer called a *conche*. Slow mixing further reduces the particle size of the cocoa solids, sugar, and milk solids and blends the ingredients. It also warms the mixture, which softens the fat layer. After many hours, the *viscosity* of the mix increases because of the decreasing particle size of the solids portion. At that point, the remainder of the cocoa butter, the lecithin, and the vanillin are added. The melted chocolate is poured into molds and allowed to solidify.

To illustrate the possible variations, some formulations for chocolate are given in Table 7-3. The quality of the products may be judged from these formulations, both from the percentage of more expensive ingredients used, as well as from the expense of longer conching treatment and finer particle size.

Baker's chocolate—An unsweetened form of chocolate that is the same as chocolate liquor. While having a bitter taste, it is used as an ingredient that adds color and chocolate flavor.

Semisweet chocolate—A form of chocolate with added sugar. The amount of sugar or sweetness can vary.

Milk chocolate—A form of chocolate that has added milk solids, sugar, and other ingredients. It is lighter in color than semisweet or unsweetened chocolate.

Conche—A mixer that slowly mixes a heated paste of chocolate ingredients to reduce the particle size and increase the thickness and smoothness.

Viscosity—The thickness of a liquid or semiliquid material.

TABLE 7-3. Basic Chocolate Coating Formulations, Showing Possible Variations

	Dark Chocolate		Milk Chocolate	
Chocolate liquor, %	35.00	15.00	17.00	10.00
Cocoa butter, %	16.95	24.20	22.00	22.70
Sugar, %	48.00	60.50	40.95	54.00
Dry whole milk, %	...	...	20.00	13.00
Lecithin, %	...	0.25	...	0.25
Vanillin, %	0.05	0.05	0.05	0.05
Fineness	5	12	4-5	12
Conche time, hr	32	8	48	8
Total fat, %	35	32	36.5	32

Chocolate chips are made by dropping melted chocolate onto cooled plates. By changing the size of the apertures through which the chocolate drips and the viscosity of the chocolate itself, chips of various sizes are obtained. The standard designation refers to the number of chips per pound; it may range from 100 (very large, usually for home baking or "gourmet" cookies) to 10,000. In wholesale baking, e.g., of chocolate chip cookies, 2,000- or 4,000-count chips are usually used.

USAGE

Tempering. Tempering ensures that the coating mixture contains only form V *seed crystals*. Proper tempering gives the cocoa butter in the mixture the desired solidity and melting characteristics. The effects of improper tempering are shown in Figure 7-3. Properly tempered chocolate coatings and compound coatings (discussed below) are resistant to the development of bloom. If the coating mixture is improperly tempered, the rapid appearance of bloom is assured. Several different tempering procedures are used (coating suppliers can provide more details).

In the complete melt method, the chocolate is warmed to 45°C (113°F) and held for about 1 hr to ensure complete destruction of all seed crystals. The mass is then cooled to 29–30°C (84–86°F) and stirred for about 10 min. It thickens as crystals form. It is then slowly warmed to a temperature just below the melting point, say 31–32°C (88–90°F), and held for several hours. Any less-stable crystals (i.e., forms II and IV) melt, and only form V seed crystals are present. The coating is then ready for use.

In partial melt methods, pretempered chocolate is warmed to 31–32°C, taking care to avoid localized "hot spots." Heating can be done with either a microwave oven or a jacketed kettle. After a holding period of a few hours, the crystal seeds are stabilized, and the coating may be used. In a variation of this, the chocolate is heated to 38°C (100°F) until it just melts. The heating is stopped, and a chunk of tempered chocolate equal to about one-fourth the weight of the melted chocolate is added. The mixture is stirred until the temperature of the mass has dropped to 31–32°C. The unmelted portion of the chunk is removed. The coating is in temper and ready to use.

Enrobing. Most products are enrobed by passing them through a "waterfall" of melted chocolate, then transporting them through a cooling tunnel to solidify the coating. Temperature control throughout this process is crucial to good quality of the final product. First, the product to be enrobed should be at 24–27°C (75–80°F). Colder product "shock cools" the chocolate, causing formation of form II and/or form IV crystals. Warmer product may melt the form V seeds, "breaking the temper" of the chocolate and promoting subsequent blooming. If the product is thick and moist (e.g., a Swiss Roll), it should stand long enough to allow internal moisture equilibration before being enrobed.

> **Seed crystals**—Small pieces of solid fat in a crystalline form that can induce the formation of additional fat crystals upon cooling.

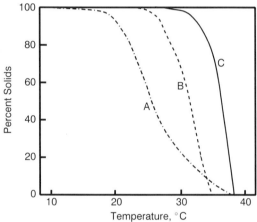

Fig. 7-3. Melting curves of cocoa butter tempered in various ways. A, tempered to give a mixture of forms II and IV. B, tempered to stabilize the form IV crystals. C, tempered to stabilize form V crystals. Since form V is the most stable, proper tempering gives the curve shown in C.

Second, cooling should not be too rapid or extreme. Ideal conditions are about 18°C (65°F) with good air circulation (to remove heat from the pieces). Too low a temperature promotes unstable form IV crystals, while warm temperatures slow the growth of form V crystals. A good temperature regime for a zoned cooling tunnel is 19–21°C (66–70°F) in the first zone, cooling to 13–15°C (55–59°F) in the middle zone, then warming again to 18–20°C (64–68°F) before the exit to prevent any condensation of moisture from the ambient air. Finally, the wrapped product should be stored at 15–21°C (59–70°F) for complete crystallization of the fat and stabilization of form V.

Moisture condensing on the coating, either from the air or from moisture migrating from the baked product, dissolves some of the sugar in the chocolate, causing stickiness and a dull appearance. In humid weather, care should be taken to prevent condensation.

The thickness of the covering is governed by the viscosity and temperature of the melted chocolate. Viscosity is determined by three factors:

- Fat content—increasing the amount of fat lowers viscosity.
- Fineness—more conching, which reduces particle size, increases viscosity.
- Emulsifiers—the addition of lecithin up to 1% lowers viscosity.

The viscosity of the coating at a set temperature (usually 40°C) is controlled by the manufacturer. In the plant, minor changes in viscosity (and hence amount of coverage) can be made by small temperature changes of the melted coating and the piece being enrobed.

EXTENDERS

Because of the cost of cocoa butter, other fats having similar solid fat index (SFI) profiles are used as partial substitutes. Palm oil, *shea oil*, and *illipé butter* are fractionated (sometimes after partial hydrogenation) to give fats that may be substituted for 10–50% of the cocoa butter. These fats have a fatty acid ratio of C18:1 (30%) to C16:0 + C18:0 (60%) that is roughly equivalent to that of cocoa butter (Table 7-4), so they cocrystallize in a stable β crystal. Figure 7-4 shows SFI curves for two such hard butters. In addition, fractions of hydrogenated cottonseed or soybean oil have been substituted for 10–25% of cocoa butter with some success.

Shea oil—A fat obtained from tree seeds grown in Africa; similar to cocoa butter in properties and composition.

Illipé butter—A fat obtained from the seeds of plants grown in tropical Asia; similar to cocoa butter in properties and composition.

TABLE 7-4. Fatty Acid Composition of Some Cocoa Butter Extenders

Fatty Acid	Cocoa Butter	Fat 1 (Palm Fraction)	Fat 2 (Palm Fraction + Shea Fraction)	Fat 3 (Shea Fraction)	Fat 4 (Palm Fraction + Shea Fraction + Illipé Butter)
C16:0	26.3	53.1	40.9	4.3	32.2
C18:0	33.8	5.9	21.3	54.3	30.2
C18:1	34.4	33.0	32.2	35.4	33.0

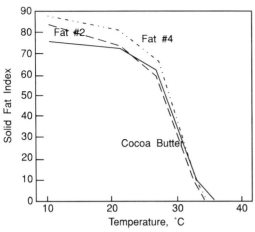

Fig. 7-4. SFI profiles of some hard butters used as cocoa butter extenders, compared to the curve for cocoa butter. The two fats (Nos. 2 and 4 in Table 7-4) are blended from palm, shea, and illipé butter.

Cocoa butter substitutes—
Hard fat sources, including hydrogenated vegetable oils, that have melting properties similar to those of cocoa butter.

Compound coatings—
Coatings containing fats other than cocoa butter but similar to regular chocolate in melting properties.

Compound Coating

Other fats are processed to give fractions with a high SFI at room temperature and melting points around body temperature. These are often called hard butters or *cocoa butter substitutes*. They are used to make coatings (*"compound coatings"*) that cost less than pure chocolate and sometimes have more desirable characteristics, i.e., melting points in the range of 38–42°C (99–108°F) for products in tropical climates. In general, these coatings have minimal compatibility with cocoa butter.

LAURIC FATS

Coconut and palm kernel oil are the base oils for these hard butters. The processes applied to their production are hydrogenation, interesterification of blends, and fractionation. Fats with a wide variety of SFI profiles and melting points may be made using one or more of these techniques. Coatings from these fats are commonly used for frozen desserts. They seem to be less subject to "cold shock" and resultant bloom formation than are chocolate and/or other types of compound coatings.

Crystal polymorphism of lauric fats is much less complex than for cocoa butter, and only a minimal amount of tempering is needed to obtain good final results. Complete melt tempering is usually used, but the holding time at 2–3 degrees C below the melting point is only an hour or so. The fatty acid composition of lauric fats is much different than that of cocoa butter, and the two kinds of fat are not at all compatible in coatings.

Specifications Common to All Good Quality Fats and Oils	
Chemical characteristics	
Peroxide value: 1 meq/kg maximum	
Free fatty acid (as oleic acid): 0.05% maximum	
Phosphorus content: 1 ppm maximum	
Physical characteristics	
Color: (Lovibond): 1.5 R, 15 Y maximum	
Flavor: Bland	
Odor: Neutral when warmed	

Hard Butter, Coating Fat Specifications				
AOM stability: 200 hr	Solid fat profiles:			
Oil stability index: 80 hr	°C	SFC	°F	SFI
Wiley mp: 38 ± 1°C	10	88 ± 5	50	64 ± 5
	15	75 ± 5	70	52 ± 4
	20	61 ± 5	80	44 ± 4
	25	47 ± 4	92	20 ± 2
	30	32 ± 4	100	6 ± 1
	35	10 ± 1	104	0
	40	0		

Nontempering coatings—
Coatings that are made of fats that do not readily undergo crystal form changes and therefore do not need tempering to inhibit chocolate bloom.

OTHER FAT SOURCES

Palm, cottonseed, and soy oils that have been hydrogenated and fractionated have been extensively investigated for use in compound coatings. A wide range of SFI profiles and melting points is available through adjustment of hydrogenation conditions and fractionation techniques. Two examples are shown in Figure 7-5. One of the interesting production techniques is to purposely "poison" the nickel catalyst by the addition of a sulfur compound such as methionine part way through the selective hydrogenation process; the resulting fat has an exaggerated, humped SFI profile. Some of the fractions from cottonseed or soy oil are partially compatible with cocoa butter and may be substituted for 10–25% of cocoa butter. (Note, however, that this is not true for the oils themselves. Admixture with soy oil— perhaps from the baked product being enrobed—may lead to rapid bloom formation in the chocolate coating.)

Coatings made with these fats are often termed *"nontempering" coatings*, because tempering is not required. In other words, the coating is simply melted, warmed to the desired application temperature, and then used. The fat is quite stable in the β crystal and shows excellent resistance to bloom formation.

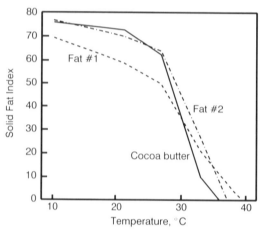

Fig. 7-5. SFI profiles of two vegetable oil hard butters compared to the curve for cocoa butter. They are blended of hydrogenated, fractionated soybean (No. 1) and cottonseed (No. 2) oils.

Troubleshooting

BLOOM

Symptom	Causes	Changes to Make
White, dusty appearance	Incorrect tempering of melted chocolate coating	Make sure recommended tempering and seeding procedures are followed closely. Check: —Melting temperature —Cooling temperature —Amount, kind of seeding material to use —Temperature after addition of seeding material Increase time in tempering unit at 31–32°C (88–90°F).
	Incorrect temperature of application	Make sure chocolate going to enrobing machine is at 31–32°C. Adjust temperature of piece to be enrobed to 24–27°C (75–80°F). Bottoming plate should be no colder than 10°C (50°F).
	Incompatible fat	Do not add lauric fats to chocolate coating. Make sure any surface fat (e.g., pan release oils) on the piece to be enrobed is compatible with chocolate.
	Incorrect storage	Store enrobed pieces at 20–25°C (68–77°F) for two days for full crystallization of cocoa butter. Do not expose to temperatures above 30°C (86°F); preferably store below 27°C (80°F).
Sugar bloom (sugar crystals on surface)	Surface moisture on piece to be enrobed	Thaw frozen pieces completely, in a low humidity atmosphere. Cooled piece temperature must be above dew point of ambient air temperature and relative humidity.

COATING APPEARANCE

Symptom	Cause	Changes to Make
Coating too thick	Coating viscosity too high	Lower viscosity by adding lecithin or fat to coating formulation.
	Coating temperature too low	Increase temperature of coating holding tank.
	Blower and shaker not set properly	Increase rate of air blower and/or shaker oscillation.
	Piece to be enrobed too cool	Decrease cooling of centers before enrobing.
	Bottom clearance too large	Decrease clearance between bottoming plate and carrier chain.
Coating too thin	Coating viscosity too low	Increase viscosity by decreasing lecithin or fat in coating formulation.
	Coating temperature too high	Decrease temperature of coating holding tank.
	Blower and shaker not set properly	Decrease rate of air blower and/or shaker oscillation.
	Piece to be enrobed too warm	Increase cooling of centers before enrobing.
	Bottom clearance too small	Increase clearance between bottoming plate and carrier chain.
Pin holes	Coating viscosity too high	Lower viscosity by adding lecithin or fat to coating formulation.
	Coating applied at too low a temperature	Increase temperature of coating holding tank.
	Excessive stirring of melted coating	Decrease agitation in holding tank.
	Piece to be enrobed is porous	Use a coating roller.
Cracking	Centers contracting in cooling tunnel	Cool centers before enrobing.
	Moisture content (wafers) too high	Bring moisture down to 3–5% before enrobing.
Greasiness	Cooling shock in cooling tunnel	Raise temperatures inside postcoating cooling tunnel.
Tails/feet on finished piece	Coating viscosity too low	Increase viscosity by decreasing lecithin or fat in coating formulation.
	Blower and shaker not set properly	Increase rate of air blower and/or shaker oscillation.
	Piece to be enrobed too warm	Cool centers to 18–24°C (65–75°F) before enrobing.

Salad Dressings

A 17th century cookbook, referring to the preparation of "salattes," admonishes the chef to "be not too generous with the vinegar, nor yet too sparing with the oil." The cruet set, a pair of containers for wine vinegar and salad oil, is still found, but pourable salad dressings today are sold preformulated, containing all the flavorings and additional ingredients in one bottle. Spoonable salad dressings, in their original concept, were not intended for salads. Mayonnaise (created for the Comte de Mayonne) was used as a sauce on seafoods such as lobster and shrimp. Today its main use is in potato, tuna, and chicken salads, in tartar sauce, and as a substitute for butter or margarine in sandwiches.

Pourable Salad Dressings

COMPOSITION

Standardized french dressing is defined in 21 CFR 169.115 as a mixture of vegetable oil (minimum 35% by weight), acidifying ingredients (vinegar, lemon juice, or lime juice), and other permissible ingredients (salt, sugars, spices, monosodium glutamate, tomato products, eggs, colorants, thickeners, citric and/or malic acids, *sequestrants*, and crystallization inhibitors). Thus a wide range of products is included, from the simplest "vinaigrette" (oil, vinegar, and a few spices) to rather complex products marketed under a variety of names.

In addition, numerous nonstandard products are available. For instance, the many "reduced-calorie" dressings contain significantly lower levels of oil. The viscosity of these products is obtained by the addition of gums, emulsifiers, and other thickeners such as microcrystalline cellulose.

Sequestrants—Compounds that bind or form complexes with a second compound so that the second compound is no longer chemically active. Positively charged metal ions such as calcium are often sequestered by compounds such as citric acid or EDTA.

OIL CHARACTERISTICS

Salad dressing oil is a refined, bleached, deodorized (RBD) oil that has been winterized. It must have a cloud point of 5.5 hr minimum, but a good-quality salad oil has a cloud point of at least 15 hr. Salad dressings are refrigerated between uses, and cloudiness is unsightly and gives a waxy mouthfeel. The addition of a crystallization inhibitor can extend the cloud point; a 5-hr oil may become a 10-hr oil, while a high-quality 15-hr oil may have its cloud point extended to 80 hr. Crystal inhibitors have a molecular structure similar to that of triglycerides but differing

Oxystearin—A preparation of oxidized and polymerized hydrogenated vegetable oil that is added to oils to inhibit the crystallization of triglycerides.

in some specific manner. They settle on the face of the growing fat microcrystals, and the difference in molecular architecture interferes with the further deposition of fat molecules. Two compounds are specifically approved by the FDA for this use: *oxystearin* and polyglycerol esters. Oxystearin is made by blowing air through cottonseed or soybean oil (hydrogenated to an IV of about 35) at 200°C (392°F). The product contains many polymeric and breakdown products, and its exact composition is unknown. Its maximum permissible concentration is 0.125% by weight. Several emulsifiers (in addition to polyglycerol esters) also inhibit crystal formation. Among those reported to do this are sucrose esters, glucose esters, and sorbitan tristearate.

Oxidative stability of the oil is particularly important, so that rancid flavors don't develop after the bottle of dressing is opened. Stability can be increased in several ways. The oil must be properly refined so that the peroxide value is less than 1 meq/kg. Citric acid is added at the end of the deodorization step at 10–50 parts per million (ppm) to sequester metal ions. Antioxidants are added at a maximum level of 0.01% (100 ppm) of any one material and 0.02% (200 ppm) of any combination. These three factors (refining, citric acid, and antioxidants) can substantially extend the active oxygen method (AOM) value of the oil and produce a dressing with keeping quality that meets the consumer's expectations. Calcium ethylenediaminetetraacetic acid (EDTA) and calcium disodium EDTA may be added to the dressing (at a maximum level of 75 ppm) to sequester metal ions contributed by other ingredients in the formula.

EMULSIFIERS AND THICKENERS

Before adding dressing to the salad greens, the user shakes the bottle to make an emulsion, so that the oil and water phases are evenly distributed. Many nonstandard dressings include the emulsifier Polysorbate 60 in their formulation at a maximum permissible level of 0.3% by weight of the total dressing to improve "home emulsification." In addition, many dressings are rather complex mixtures containing dairy products (e.g., buttermilk powder, blue cheese), spices (e.g., ground mustard, turmeric powder), dried vegetable pieces (e.g., onion, green and red peppers), and other items according to the creativity of the product developer. An emulsion stabilizer helps maintain uniformity of this mixture and increases the consumer appeal of the product.

The viscosity of a simple emulsion of 35% oil in water is almost the same as the viscosity of water. For many dressings, a higher viscosity is desired so that the dressing does not drain quickly from the salad greens and collect in the bottom of the salad bowl. Soluble gums are used at a concentration of 0.05–0.3% to give the desired viscosity. Some of the names commonly seen on dressing labels include: polyethylene glycol alginate, xanthan gum, modified cellulose gum, and carrageenan. (This list is not exhaustive.) The main requirement for the gum (besides being able to contribute the desired viscosity) is that it must be stable in an acidic environment; certain gums slowly hydrolyze at low pH, with a resultant decrease in viscosity.

Another thickener is microcrystalline cellulose, used at 1–2% concentration. The water adsorption capacity of this material is quite high, and it increases viscosity by decreasing the amount of continuous (water) phase in the formulation. It is most often seen in dressings that contain tomato products, where it gives a smooth texture and imparts a certain degree of *thixotropy* to the mixture so that the dressing clings to the surface of the salad greens.

Thixotropy—A physical property that makes a gel or semisolid turn into a liquid and flow when stirred.

Spoonable Salad Dressings

MAYONNAISE

Composition. Standardized mayonnaise (21 CFR 169.140) contains vegetable oil (minimum 65% by weight), acidifiers (vinegar, lemon juice, or lime juice), egg yolks (liquid, frozen, or as the whole egg), and other ingredients (salt, sugars, spices, monosodium glutamate, sequestrants, citric and/or malic acid, and crystallization inhibitors). Mayonnaise is an oil-in-water emulsion, stabilized by the lipoprotein components of egg yolk. While the legal minimum for oil content is 65%, mayonnaise at this oil level is rather thin (has low viscosity); the usual commercial product today contains 77–82% oil. Liquid egg yolk (45% solids) is the emulsifier, at 5.3–5.8% of the total formula weight. The other usual ingredients are 100-*grain* vinegar (2.8–4.5%), salt (1.2–1.8%), sugar (1.0–2.5%), and mustard flour (0.2–0.8%). Sometimes whole eggs (25% solids) are substituted for egg yolks on a total solids basis (i.e., whole eggs at 9.5–10.4% of the total formula weight); this gives a somewhat "stiffer" product than egg yolks. Paprika oleoresin may be used if a darker yellow color is desired, and garlic, onion, or other spices may be added. Water is added to bring the total formula weight to 100%.

Grain (vinegar)—A measure of the strength of vinegar in which 10 grains is equal to 1% acetic acid.

Oil is the dispersed phase in the emulsion. If all the droplets are spherical, incompressible, and have the same diameter, the maximum volume percentage of oil is 72% (Fig. 8-1a). (For example, if a 1,000-ml container were filled with small, uniform ball bearings, then 280 ml of water would be needed to fill the space between the bearings.) If the droplets differ in diameter, then small droplets can fill in the spaces between larger drops (Fig. 8-1b). This is the case with mayonnaise; otherwise, when

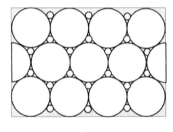

Fig. 8-1. Packing of the internal phase in an emulsion. a, All the droplets are spherical, incompressible, and have the same diameter. The internal phase can occupy, at most, 72% of the volume. b, Some droplets have a smaller diameter and can fit in between the larger droplets. The internal phase can occupy more than 72% of the volume.

TABLE 8-1. Relationship Between Weight Percent and Volume Percent

Weight Percent	Volume Percent
65	67.1
77	78.6
82	83.3

Inverted emulsion—An emulsion in which the continuous phase becomes the dispersed phase and the dispersed phase becomes continuous.

Rheological—Describing flow properties caused by an outside force or deformation.

Bingham plastic—A material that behaves like a solid until it is stirred at a high enough rate so that it starts to flow and behave like a liquid.

Shear stress—The application of a force, such as stirring, to a material.

Yield value—The amount of shear or stirring needed to turn a Bingham plastic from a solid to a liquid.

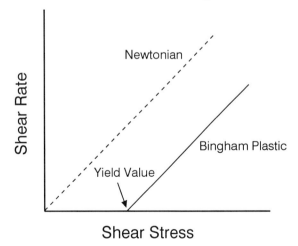

Fig. 8-2. Rheology of mayonnaise. The rate of flow of a Newtonian fluid such as water is directly proportional to the force applied (shear stress). A Bingham plastic does not flow until the shear stress exceeds the yield value; then the flow rate is directly proportional to the additional shear stress.

the oil volume percentage exceeded 72%, the emulsion would *invert* (the oil droplets would coalesce and the emulsion would become a water-in-oil emulsion). Note that the proportion of oil in the emulsion must be given as volume percentage; because the density of oil is about 0.91 g/ml, there is a difference between weight percentage and volume percentage (Table 8-1).

This "filling in the spaces" also has a major effect on the *rheological* (flow) characteristics of the product. Mayonnaise is a "*Bingham plastic*" in rheological terms (Fig. 8-2). When *shear stress* (stirring) is applied at low levels, mayonnaise acts like a solid; that is, it does not flow, although it may fracture (break). At some level of shear force, called the yield value, the mayonnaise begins to act like a liquid. Put another way, if a mass of mayonnaise is simply cut, the vertical edge does not flow, because the shear stress of gravity is less than the *yield value*. However, it can be spread smoothly on a slice of bread, because the shear force applied by the knife blade is greater than the yield value. The size of the yield value is related to the amount of dispersed phase (oil) in excess of the theoretical limit of 72% (by volume). Thus, a mayonnaise with the legal minimum of 65% oil by weight has a very low yield value and is seen by users as being too thin. Mayonnaise with 80–84% oil by weight has a high yield value and is seen as dry or rubbery by home consumers, although it is preferred for institutional use because it does not soak into bread in sandwiches or soften and flow over salads.

As mentioned above, the lipoproteins in egg yolk are the main emulsifying agents in mayonnaise. They are partially denatured by the low pH of the vinegar used, and their emulsification capabilities are improved. If whole eggs are used, the egg white albumins are also denatured by low pH and have some emulsifying properties. The denatured albumins give the aqueous phase a somewhat meringue-like nature; the resulting mayonnaise may be slightly fluffier than mayonnaise made only with yolk, and the yield value is slightly greater. The choice between the two types of product is mainly a matter of individual consumer preference.

An oil with a long cloud point time is essential for mayonnaise. Storage, after the jar is opened, is in the refrigerator. If fat crystals form, the balance of emulsion-stabilizing forces is disrupted and the emulsion breaks. The practice of winterization was first applied to cottonseed oil intended for making mayonnaise in the early days of its commercial manufacture. The factors mentioned earlier apply here; the oil processor must do a good job of winterizing the oil, and crystal inhibitors may be used to further guard against clouding.

Manufacture. The preparation of mayonnaise is difficult, and science is less important than experience, skill, and art on the part of the equipment operators in making an acceptable final product. Nevertheless there are certain basic principles that apply. Understanding them can help to solve production problems when they arise.

The traditional method of mayonnaise production uses a *planetary mixer* (e.g., a Hobart) with a paddle. The yolk and other dry ingredients are blended; then the oil is added while mixing continues, slowly at first, then more rapidly when the mass begins to thicken (without allowing any puddles of oil to form). After all the oil is incorporated, the vinegar is added and the mixture is thoroughly blended. This method is used in homes and gourmet restaurants that have their own special blends.

Commercial manufacture is a two-step process. A premix is made, in a manner similar to the traditional method. Often mayonnaise from a previous batch is placed in the mixer to give a high-viscosity medium into which the mixer blades can readily disperse the oil. The egg and dry ingredients are added; mixing is initiated; and the oil and vinegar are pumped in at a controlled rate (the flow of vinegar is not completed until after all the oil is in). The viscosity of the premix depends in part upon the characteristics of the particular batch of egg being used and may be varied by changing the timing and rate of the oil and vinegar addition. This rather soft premix is then pumped to a *colloid mill*, of which several different types are available. The purpose of this step is to reduce the diameter of the oil droplets. The viscosity of the final product depends upon the degree of subdivision attained in the mill; this in turn depends upon the mill settings (clearance between rotor and stator, speed of the rotor) and the premix viscosity. The ideal is to obtain a high degree of subdivision but with a large variation in droplet diameters so that the situation shown in Figure 8-1b results. If too much subdivision occurs, so that all droplet diameters approach the theoretical minimum (governed by interfacial tension), then variable space filling is lost, and the situation shown in Figure 8-1a is approached. In this case, since the volume percentage of oil is greater than 72%, the emulsion inverts. The ability of the operator to adjust the mill to achieve the proper degree of subdivision depends, as stated earlier, more on skill and art than on science. With the Dixie-Charlotte mill, the clearance between the rotor and stator is usually in the range of 0.025–0.040 inches (0.6–1 mm).

Planetary mixer—A mixer with a bowl and paddle arrangement that uses circular motion.

Colloid mill—A high-speed grinder that reduces the size of the particles to a fine dispersion.

STARCH-BASED DRESSINGS

Spoonable salad dressing was first developed as a low-cost alternative to mayonnaise. Today it is accepted as a somewhat different product, purchased and judged (by consumers) on its own merits. While it is similar to mayonnaise in rheology (it is a Bingham plastic, and its yield value is a factor in acceptability), these characteristics are dependent on the nature of the food starch used, not the state of emulsification of the oil.

Composition. Standardized salad dressing (21 CFR 169.150) contains vegetable oil (minimum 30% by weight), egg yolk (minimum 4% by

Cross-linked starch—Starch that is chemically modified by linking the starch molecule chains together laterally with a chemical reagent.

Waxy maize starch—Corn starch that is high in the amylopectin type of starch and low in the amylose type of starch.

weight), acidifiers (vinegar, lemon juice, or lime juice), a paste prepared from a suitable food starch, and other ingredients (salt, sugar, spice, monosodium glutamate, thickeners, citric and/or malic acid, sequestrants, and crystallization inhibitors). A typical commercial product might contain (by weight) 35% salad oil, 4% liquid egg yolk, 2% salt, 11% sugar (or 15.5% high-fructose corn syrup), 0.5% mustard flour, 11% vinegar (100-grain), 6% modified starch, spices, and water to total 100%.

The starch is the key component in this formulation because it provides the desired structure as well as a creamy texture in the finished product. The best type of starch to use is a highly *cross-linked*, stabilized, *waxy maize starch*. Waxy maize is essentially 100% amylopectin. Cross-linking with sodium trimetaphosphate creates phosphate diester linkages within the amylopectin. This starch requires rather high temperatures for gelatinization, then sets up into a soft gel upon cooling. The cross-linked starch is stable against hydrolysis in the low-pH environment of the finished dressing; without this stability, the starch gel would soften during storage. Stabilization of the starch is accomplished by attaching a substituent to the small side branches of the amylopectin molecule. The recommended derivative is a hydroxypropyl moiety, obtained by treating the starch with propylene epoxide. This stabilization interferes with recrystallization of the side chains and therefore maintains a creamy texture in the finished dressing during storage, particularly if the dressing is held at refrigerator temperatures or even frozen.

Manufacture. Spoonable salad dressing is made in two stages. First, the oil is emulsified with the egg yolk. Meanwhile, the starch is gelatinized by heating with the water; after cooling, the vinegar, sugar, salt, spices, etc. are mixed into this paste. The cooled paste is then fed into the premixer and blended into the oil-egg emulsion. This soft mass is then pumped to a colloid mill, where the final high viscosity and smooth texture are generated. The extent of oil subdivision is generally less in salad dressing than in mayonnaise, i.e., the colloid mill settings are somewhat more open (e.g., 0.045–0.050 in. [1.14.–1.27mm] clearance in the Dixie-Charlotte mill). The gelatinization operation is the crucial step in attaining good final-product quality, and the details of time and temperature depend to a large degree upon the specific starch being used. A full discussion is outside the scope of this book; the starch supplier should be the first line of information for the proper operation of the starch cooker.

Troubleshooting

MAYONNAISE

Symptom	Causes	Changes to Make
Emulsion breakdown	Crystal formation in the oil	Use a winterized oil with a longer cloud point time.
	Excessive subdivision of oil droplets	Open clearance on colloid mill slightly. Decrease viscosity of premix emulsion.
Rancid flavor	Oxidation of oil	Store oil under N_2 before use in manufacture. Package product under N_2. Use a lightly hydrogenated, winterized oil.

SALAD DRESSING

Symptom	Causes	Changes to Make
Loss of "body"	Acid hydrolysis of starch	Use a highly modified waxy maize starch. Add vinegar after starch has been gelatinized and cooled.
	Low gel viscosity of starch	Use a modified starch with a higher cold viscosity.
	Freeze-thaw breakdown of starch gel	Use a stabilized starch for freeze-thaw resistance.
Rancid flavor	Oxidation of oil	Store oil under N_2 before use in manufacture. Package product under N_2. Use a lightly hydrogenated, winterized oil.

Nutritional Topics

Lipids are an important part of animal (including human) nutrition. Excessive dietary intake of lipids (e.g., fats or cholesterol) can lead to health problems, and numerous governmental and professional groups recommend that fat intake be limited so that it provides no more than 30% of the total *calories* consumed each day. However, certain groups appear to advocate a zero level of lipid consumption, on the mistaken assumption that this can only improve our health. In this chapter, some of the contributions of lipids to good health are briefly discussed. It is important to recognize that research on this subject continues, and the last word has by no means been spoken. If the reader requires more detailed or up-to-date information, the extensive literature on nutritional aspects of fat and cholesterol intake should be consulted. Moreover, the guideline "Moderation in all things" is as worthwhile today as it was when it was first given many centuries ago. Extremes of any sort, including dietary manipulation, are not helpful and may well be damaging.

Role of Dietary Fats

CALORIES

Fat is a high-density source of calories. It contributes nine *kilocalories* of *energy* per gram, whereas starch or protein contribute four kilocalories per gram. Through most of human history, when obtaining adequate food energy was sometimes a problem, fatty animal tissue and oil from vegetable sources were prized components of the diet. Animal *metabolism* reflects this evolutionary fact; when excess energy is available, the body converts it to fat and stores it in *adipose tissue* for use when energy intake is inadequate. The modern food industry (consisting of production, processing, and distribution systems) has negated "lean times" as a seasonal fact of life for most people in developed countries; adequate food energy is available year round. Now many people are more worried about *obesity* than about lack of food. However, obesity is not due simply to excess intake of fat; when more calories from any source (i.e., starch, protein, and fat) are consumed than are expended in exercise and metabolic processes, the body obeys its inherited programming and stores fat.

From 1895 to 1905, W. O. Atwater, a chemist at the Agricultural Experiment Station in Storrs, Connecticut, measured the heat content of a

Lipids—A class of compounds found in nature that are soluble in organic (nonpolar) solvents such as ether or hexane. Triglycerides, cholesterol, and vitamin A are examples.

Calories—The energy contained in food components that gets released and absorbed by the body during metabolism. One calorie is the amount of energy required to raise the temperature of 1 g of water by 1°C.

Kilocalories—1,000 calories, which is the amount of energy required to raise the temperature of 1 kg of water 1°C. When referring to the caloric content of the diet, the *kilo* part is often dropped and just the term *calories* is used, usually written with a capital C. In the statement that fat has 9 Calories per gram, it means *kilocalories*.

Energy—The value derived from the digestion and metabolism of food components that is turned into work in a body or tissue.

Metabolism—A series of reactions in the body by which complex food molecules are degraded to simpler forms, energy is extracted, and complex molecules are reformed by the body.

Adipose tissue—Tissue in which fat is stored.

Obesity—The condition of being overweight due to the accumulation of fatty tissue in the body. Obesity usually is considered being more than 20% over ideal weight.

Bomb calorimeter—An instrument that burns a sample of food and determines how much heat energy it releases.

Gut—General term for the digestive tract or intestinal tract.

Available heat—The caloric or energy content of food materials, taking into account the amount of material that is absorbed from the digestive tract.

wide variety of foods and food components using a *bomb calorimeter*. This analytical method determines the amount of heat released when the sample is burned completely in an atmosphere of pure oxygen. Atwater also conducted feeding studies with human volunteers, assessing what fraction of each of the foods they ate was absorbed in the *gut*. Multiplying the analytical heat content by the fraction absorbed, he obtained values for the *available heat* content of many foods. The numbers vary over a range of 0.5 kilocalories per gram for the foods in each group (carbohydrates, proteins, and lipids), but the average for each group has been accepted as the well-known Atwater values of 4, 4, and 9 Cal/g for carbohydrates, proteins, and fats, respectively. The term Cal ("big C" calories), which is used in nutritional labelling, is approximately equivalent to kcal (kilocalories), the correct scientific designation. It is convenient to reserve Cal for the results of calculations using the Atwater factors (with their inherent approximations) and kcal for more exact scientific calculations.

MEMBRANE STRUCTURE

Bilayer lipid structures are important at the cellular level in animals. Cell membranes consist of a double layer of phospholipid, with embedded proteins that transport molecules and ions across the cell wall (Fig. 9-1). The hydrophilic end of the phospholipid is located either at the inner or the outer surface of the cell wall, and the fatty acid chains of the phospholipids form the structure of the membrane. This bilayer structure is found not only in cell walls but also in the cell interior (in microsomes, the reticular endothelium, etc.), as well as in the myelin nerve sheath, eye retina, and other structures.

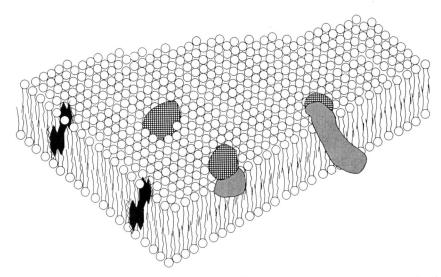

Fig. 9-1. The structure of cell membranes. The main component is phospholipids, which have their fatty acid chains in the interior of the membrane and their charged end groups on the surface. Some cholesterol (solid back structures) is found inserted in the membrane, and proteins (cross-hatched and shaded structures) performing various functions are attached to one side or the other or else penetrate the membrane.

This membrane must have a certain degree of fluidity; if it is too soft, the cell wall is disrupted, but if it is too hard, the protein chains can't move to perform their transport function. The fluidity depends upon the degree of unsaturation of the fatty acids. Very roughly, the fatty acid composition of the cell wall phospholipid is 40% saturated (in C16:0 and C18:0), 20% monounsaturated, and 40% polyunsaturated. Most of the diglyceride portions contain one saturated and one unsaturated fatty acid; generally only a trace of disaturated diglyceride is found, and about 20% of the diglycerides contain no saturated acid.

The phospholipids are synthesized in the body rather than obtained from dietary sources. Fatty acids are esterified to glycerol to make a diglyceride, which is then phosphorylated and reacted with another compound (i.e., ethanolamine, choline, inositol, or serine) to make the final phospholipid. While most of the fatty acids required for making phospholipids (with the exception of the *essential fatty acids* [see below]) can be synthesized in the body, this is a relatively inefficient way of obtaining them. Fatty acids absorbed from the gut can be used directly to make phospholipids, and if the composition of dietary fat roughly matches that of the membrane phospholipids, the entire process is more efficient.

Essential fatty acids—A class of fatty acids required in the diet of humans. Linoleic and α-linolenic acids are essential fatty acids.

VITAMINS

Vitamins A, D, E, and K (Fig. 9-2) are soluble in fat and are associated with the lipid portion of foods. A well-balanced diet normally provides adequate vitamin intake. The main reasons for deficiency are: 1) inadequate dietary supply and 2) excessive excretion as a result of high fat content in stools ("steatorrhea," which is associated with impaired fat digestion).

Vitamin A is formed by oxidative cleavage of *carotenes*. β-Carotene, for example, gives rise to two molecules of vitamin A (for the structure

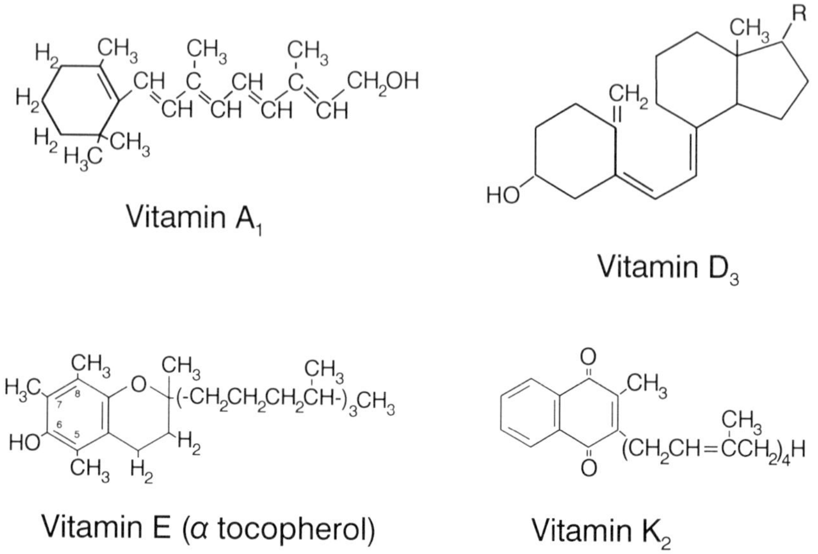

Carotenes—A class of fat soluble compounds that are yellow to red in color. Some carotenes, such as β-carotene, are converted to vitamin A in the body.

Fig. 9-2. The fat-soluble vitamins A, D, E, and K.

of β-carotene, imagine two molecules of vitamin A with the side chains linked end-to-end and reduced). Other *carotenoids* are not symmetrical, and cleavage produces only one molecule of vitamin A. (It should be noted that not all carotenoids are precursors of vitamin A.) The main deficiency symptom is loss of night vision, but extreme deficiency can lead to blindness. Fruits and vegetables supply two-thirds of our dietary vitamin A, through conversion of carotenoids. Fish liver oils are the most potent natural source of vitamin A.

Vitamin D is produced (for the most part) by irradiation of cholesterol (compare its structure in Figure 9-2 with that of cholesterol in Figure 9-4). The second ring is first dehydrogenated to give 7-dehydrocholesterol; then the ring is opened to form vitamin D_3. Other slight variants on this structure generate other members of the vitamin D group. Since irradiation is needed for its formation, vitamin D is sometimes referred to as "the sunshine vitamin." The main deficiency symptom is rickets—improper bone development in children who are exposed to little sunlight. Ultraviolet irradiation of diets is one solution, as is more exposure to sunlight. The intermediate, 7-dehydrocholesterol, is found in (relatively) high amounts in skin. Fish oil also contains members of the vitamin D group; the metabolic route leading to their formation is not known.

Vitamin E is also known as *tocopherol*. It has several isomers that have different distributions of the methyl groups and different vitamin activities, expressed as *International Units* (IU) per milligram of isomer (Table 9-1). The main deficiency symptom is loss of fertility. Rats grown exclusively on cow's milk are unable to bear young; the fetus dies three to four days after conception and is resorbed. Young animals grown on vitamin E-deficient diets develop muscular dystrophy. The corrective property of vitamin E appears to be linked to its antioxidant characteristics; tocopherols are known to inhibit autoxidation of polyunsaturated fatty acids. The best natural sources of vitamin E are vegetable oils. Wheat germ oil is especially rich in vitamin E.

Vitamin K is involved with blood clotting reactions. In deficient animals, a relatively minor injury may lead to life-threatening blood loss. The main source of vitamin K is the bacterial flora found in the large intestine. These bacteria synthesize the vitamin, which is then absorbed into the body. Other than steatorrhea, the main cause of vitamin K deficiency is a regimen of antibiotics that markedly inhibits growth of intestinal bacteria.

TABLE 9-1. Tocopherol Isomers

Isomer	Structure	IU of Vitamin E per Mg of Isomer
α-tocopherol	5,7,8-trimethyl	1.49
β-tocopherol	5,8-dimethyl	0.60
γ-tocopherol	7,8-dimethyl	0.19
δ-tocopherol	8-methyl	0.01

ESSENTIAL FATTY ACIDS, PROSTAGLANDINS

Studies in the late 1920s showed that rats on a fat-free diet developed severe symptoms: loss of hair, lesions in the tail, sterility, and kidney damage. The effects could be reversed by feeding ω6 polyunsaturated fatty acids (linoleic or arachidonic acid). Similar consequences were seen with mice, dogs, and pigs. The syndrome is called essential fatty acid (EFA) deficiency. Animals have little or no ability to synthesize EFAs from other fatty acids, such as stearic or oleic, so these fatty acids must be provided by the diet. EFAs are *precursors* for prostaglandins, which in turn exert a broad spectrum of physiological effects on body tissues. Deficiency in these materials causes the symptoms observed. EFA deficiency is seldom seen in humans because our diet is usually sufficiently varied to supply the required polyunsaturated fatty acids. The recommended minimum intake of EFAs has been set at 2% of dietary calories in the form of linoleic acid and 0.5% of dietary calories in the form of linolenic acid.

In mammals, synthesis of fatty acids other than EFAs involves two processes: 1) chain elongation by addition of two carbon units to the carboxyl end of the chain and 2) desaturation somewhere between carbons 3 and 10. For example, linoleic acid (which has double bonds at positions 9 and 12) is desaturated to C18:3 ω6 (γ-linolenic acid, with double bonds at positions 6, 9, and 12). Elongation forms homo-γ-linolenic acid (C20:3 ω6), and desaturation between carbons 5 and 6 gives arachidonic acid (C20:4 ω6). Arachidonic and homo-γ-linolenic acids react further to form the two main families of prostaglandins (Fig. 9-3). Eicosapentanoic acid (C20:5 ω3, derived from linolenic acid) can also be converted to another family of prostaglandins. The importance of the different prostaglandin species in producing various physiological effects is still not clear.

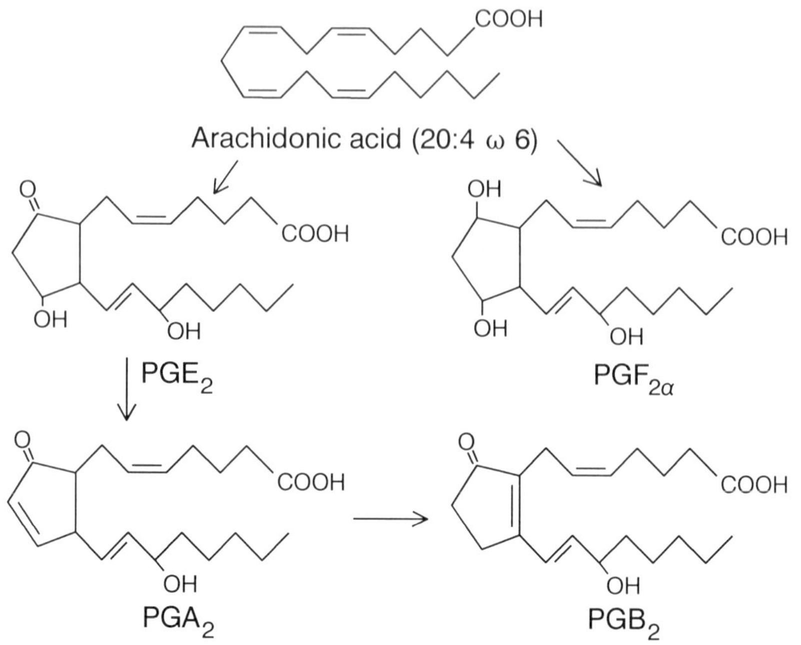

Precursors—Compounds that are the basis for the formation of other compounds.

Fig. 9-3. Prostaglandins are formed from polyunsaturated fatty acids. Arachidonic acid is first peroxidized (at position 15) by a tissue lipoxygenase; then this arachidonic hydroperoxide undergoes a series of further reactions to form PGE$_2$, PGA$_2$, and PGB$_2$. PGF$_2$ is on a separate reaction path. Prostaglandin families from other polyunsaturated fatty acids are designated by different subscripts; those from homo-γ-linolenic acid are PGE$_1$ etc., while those from eicosapentenoic acid (C20:5 ω3) are PGE$_3$ etc.

Cholesterol and Atherosclerosis

Cholesterol deposits on the walls of arteries, forming *plaques* that continue to thicken and constrict the flow of blood and eventually shutting it off completely. The results are high blood pressure and damage to the organs with blocked blood supply. Much effort is expended to deal with this medical condition and to try to minimize it by dietary changes. Cholesterol has become a "bad actor" in current views, but if it is so damaging to health, why does the body contain so much of it and continue to synthesize more? The answer, of course, is that cholesterol and other materials derived from it play many significant roles in maintaining good health. The problem arises when levels of cholesterol in *serum* rise so high that normal metabolic processes can't handle it.

A distinction should be made between dietary cholesterol and serum cholesterol. Dietary cholesterol is cholesterol that is consumed as a component of dietary lipids derived from animal sources. Reported concentrations of cholesterol in certain lipids are: butter, 0.24–0.50%; lard, 0.11–0.12%; and beef tallow, 0.08–0.14%.

Serum cholesterol is found in the blood serum lipids (see below). The concentration of serum cholesterol is determined by the balance between a number of processes:

- Synthesis de novo in the body
- Absorption from the contents of the gut
- Conversion to bile acids, vitamin D, and steroid hormones
- Excretion into the intestinal lumen
- Inclusion in cell membranes
- Deposition on arterial walls.

The first two factors, of course, increase serum cholesterol, while the other processes lower it. All except the last are under some form of feedback control. For example, high serum cholesterol levels inhibit de novo synthesis. A high concentration of dietary cholesterol favors absorption from the gut, while a low concentration favors excretion into the gut. Also, the nature of the lipoprotein that contains the cholesterol (see below) influences the rate at which one or more of these reactions occurs. For example, high-density lipoprotein (*HDL*) cholesterol is the form that is transported to the liver for conversion, whereas low-density lipoprotein (*LDL*) cholesterol is the form that deposits on arterial walls.

In summary, controlling serum cholesterol levels is an extremely complicated process, which is poorly understood at this time. Changes in the diet influence it, but the exact mechanisms by which serum cholesterol is raised or lowered are still under investigation.

POSITIVE CONTRIBUTIONS OF CHOLESTEROL

Cholesterol is synthesized by a series of condensations of lipid materials, starting with *acetylCoA*. The result is a condensed four-ring structure bearing an eight-carbon side chain (Fig. 9-4a). Normally about two-thirds of serum cholesterol is esterified to a fatty acid. Many cell membranes contain some cholesterol ester, which is apparently necessary for proper membrane fluidity and perhaps plays a role in certain cross-

Plaque—A deposit on the inner walls of arteries that grows and eventually causes blockage.

Serum—The liquid fraction of the blood that remains after clotting. Cholesterol is measured in the serum portion of the blood.

HDL—High-density lipoproteins. Molecular complexes found in the blood that carry cholesterol. Cholesterol bound to HDL is being transported and is considered a good type of cholesterol

LDL—Low-density lipoproteins. Molecular complexes found in the blood that attach to cholesterol. Cholesterol bound to LDL is considered bad cholesterol because it deposits on the walls of arteries.

AcetylCoA—A metabolic form of carbon in the body that is a basic building block for the formation of more complex molecules.

membrane transport reactions. Serum cholesterol (including esters) is suspended in a variety of lipoprotein complexes (see below).

In the liver, cholesterol is converted to bile acids (Fig. 9-4b), which are excreted into the gut. There they act as surfactants, emulsifying and suspending fats for digestion. An inadequate level of bile acids contributes to steatorrhea (fatty stools), leading to deficiencies in fat-soluble vitamins. Bile acids also help suspend cholesterol esters in the bile duct, and if this suspending action fails, hard lipid crystals form, blocking the duct and causing great pain. Finally, there is evidence that bile acids suspend cholesterol in the gut, decreasing its absorption through the intestinal wall and increasing its excretion, thus aiding in lowering serum cholesterol levels.

As mentioned above, cholesterol is irradiated in the outer layers of skin, where it forms vitamin D, which is vital to proper bone growth and skeleton development.

Many hormones are derived from cholesterol. To mention merely a few examples, there are:

• Progesterone, an adrenal steroid of endocrinological importance (Fig. 9-4c)
• Testosterone, an androgen (male sex hormone) (Fig. 9-4d)
• Estradiol-17β, an estrogen (female sex hormone) (Fig. 9-4e).

Numerous other hormones are derived from cholesterol. A good college biochemistry text can supply more extensive information.

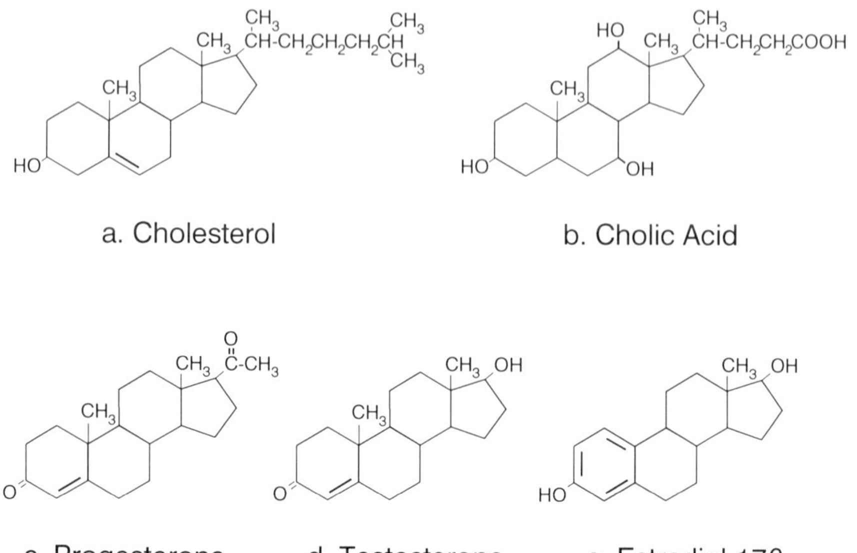

a. Cholesterol

b. Cholic Acid

c. Progesterone   d. Testosterone   e. Estradiol-17β

Fig. 9-4. Cholesterol and some physiologically important derivatives. a, Cholesterol. About 65% of the hydroxyl group is esterified to a fatty acid. b, A bile acid; the carboxyl group is often linked to glycine or taurine, forming even more powerful detergents. c, Progesterone, a hormone made in the adrenal cortex and having widespread effects in the endocrine gland system. d, Testosterone, the basic male sex hormone; other androgens are derived from this form. e, Estradiol-17β, the basic female sex hormone; other estrogens are derived from this molecule.

BLOOD SERUM LIPIDS

Serum lipoproteins. Lipids in blood are emulsified and stabilized by phospholipids and proteins. The total complex is termed *"lipoproteins."* Clinical analysis of blood serum lipids generally gives two types of data: the total concentration of serum cholesterol (combining free and esterified cholesterol), and quantitation of the main families of lipoprotein. The latter is obtained by analytical ultracentrifugation—the groups are separated by their sedimentation or flotation in salt solutions and are characterized by their densities (Table 9-2).

Chylomicra are essentially droplets of emulsified fat. They are formed by the transport of fat across the *intestinal mucosa* and are carried in the blood to adipose tissue or to other organs where the fat is metabolized. The chylomicron concentration is high shortly after a meal and tends to drop fairly quickly. Chylomicra have little to do with *atherosclerosis*.

Clinical reports and research on effects of diet on serum cholesterol level generally focus on two lipoprotein groups—high-density lipoprotein (HDL) and low-density lipoprotein (LDL). Put simply, HDL is "good" lipoprotein and LDL is "bad" lipoprotein. The difference has to do with how they function metabolically. HDL is a dedicated carrier for cholesterol, delivering it to the liver and similar organs, where it is converted into one of the other desirable forms (e.g., bile acids). LDL, on the other hand, seems to provide primarily a parking place for cholesterol. When, on its journey through the circulatory system, LDL comes into contact with cholesterol plaque (or an injury site on the blood vessel wall where plaque deposition may start), it releases some of its load of cholesterol, thus increasing plaque growth. The variation in behavior of the two classes of lipoprotein is probably related to the nature of the protein involved with each. The proteins found in each of the classes are different, and those differences influence how they interact with receptors and other organs in the body.

Influence of dietary fatty acids. Much effort has been put into understanding how different types of fatty acids influence serum cholesterol

---

Lipoproteins—Compounds that contain a protein molecule attached to a lipid molecule.

Chylomicra—An emulsified form of fat developed in the intestine to transport the absorbed fat in the blood to other organs.

Intestinal mucosa—Tissue on the walls of the intestine that controls the transport of digested material into the blood stream.

Atherosclerosis—A disease of the circulatory system caused by clogging of the arteries.

TABLE 9-2. Blood Serum Lipoproteins

	Chylomicra	Very Low Density (VLDL)	Low Density (LDL)	High Density (HDL)	Very High Density (VHDL)
Density	<1.006	1.006–1.019	1.019–1.063	1.063–1.21	>1.21
mg/100 ml plasma	100–250	130–200	210–400	50–130	290–400
Composition, approximate percentage					
Protein	2	9	21	33	57
Phospholipids	7	18	22	29	21
Cholesterol, free	2	7	8	7	3
Cholesterol ester	6	15	38	23	14
Triglyceride	83	50	10	8	5
Fatty acids	…	1	1	…	…

level. The effects have been considered from two standpoints: influence on total serum cholesterol and influence on the levels of HDL and LDL. As with any active field of research (especially medical-related research), there are disparities in opinions. The best current consensus is summarized in Table 9-3. Saturated and *trans*-unsaturated fatty acids lower HDL and elevate LDL and total cholesterol. Of the saturated fatty acids, myristic (C14) is the most *hypercholesterolemic*; palmitic (C16) is somewhat less so, and lauric and stearic (C12 and C18) are mildly hypercholesterolemic. Polyunsaturated fatty acids lower all the categories. While this is good in terms of total cholesterol, the decrease in the "good" HDL is not desirable. Monounsaturated fatty acids have the most favorable influence, raising HDL, lowering LDL, and lowering total cholesterol. The medium- and short-chain fatty acids (butyric through capric acids) are neutral with respect to serum cholesterol metabolism. Based on this simplified summary of extensive, complex data, the current emphasis on lowering saturated fat and increasing monounsaturated fat in the diet makes sense for people who are (or may become) hypercholesterolemic.

Hypercholesterolemic— Describing those dietary components that tend to raise the level of serum cholesterol.

trans **Fatty acids.** Hydrogenation of vegetable oils converts some fraction of the double bonds from the *cis* to the *trans* isomer (see the discussion of hydrogenation in Chapter 4). The daily intake of *trans* fatty acids in the current U.S. diet has been estimated at 8 g/day. Some studies, using diets containing three to four times this level of *trans* fatty acids, found some elevation of LDL cholesterol and lowering of HDL cholesterol, as compared with the effects of a similar diet in which most fat calories came from oleic acid. The effects of *trans* fatty acids (on HDL and LDL levels) appear to be about half as great as the changes brought about by equivalent amounts of saturated fatty acids. Thus, while this factor should be kept in mind when making dietary adjustments, it is certainly not grounds for eliminating hydrogenated fats (e.g., margarine) from the diet.

At 8 g per day, *trans* fatty acids contribute about 3.5% of daily caloric intake. The recommended maximum level of calories from saturated fat is 10%. It has been suggested that *trans* fatty acids should be included

TABLE 9-3. Influence of Fatty Acids on Blood Cholesterol Classes

Fatty Acid Type	Effect on		
	Total Serum Cholesterol	HDL[a] Level	LDL[b] Level
Saturated (C12– C18)	Raise	Lower	Raise
Monounsaturated	Lower	Raise	Lower
Polyunsaturated	Lower	Lower	Lower
trans-Unsaturated	Raise	Lower	Raise
Short, medium (C4–C10)	...[c]	...	...

[a] High-density lipoprotein.
[b] Low-density lipoprotein.
[c] Neutral or none.

with saturated fat in preparing nutritional labels for foods. Such a move would require regulatory changes by the Food and Drug Administration, and (as of this writing) no such action has been initiated. At the present time, *trans* fatty acids are included in the total fat category on nutritional labels but are not included in the numbers for saturated fats and are also not included in monounsaturated and polyunsaturated fats (which must be *cis* isomers, by definition).

Polyunsaturated Fats and Cancer

The age-adjusted mortality rate from cancer in the United States has increased by about 8% over the last 30 years. Most of this increase is from lung cancer, caused primarily by cigarette smoking. The incidence of stomach cancer has decreased during this same period, while age-adjusted mortality rates for nearly all other cancer types have been essentially unchanged or have decreased slightly. Suggestions that dietary fat may in some way be connected with human cancer are based on *epidemiological data*; death rates in different countries from certain types of cancer appear to be correlated with total fat intake or intake of animal fat and/or cholesterol. Often, however, this correlation does not hold up when segments of the population within a single country are examined. Overall, there is no convincing evidence for a simple causal relationship between dietary fat and cancer.

Animal studies have found that an elevated level of fat in the diet increases the response to chemically induced carcinogenesis. In other words, high levels of fat intake may act as a promoter of cancer but not an initiator. This has been noticed, in particular, with breast and colon cancers in the animals. There has been some indication that this promoting activity may be more evident with polyunsaturated fats, as compared to saturated or monounsaturated fats. The hypothesis is that fatty acid peroxides (or hydroperoxides) are the promoter species. This idea receives some support from findings that natural antioxidants (vitamins A, C, and E) appear to have anticarcinogenic activity.

While the focus has been on fat content of the diet as a factor in *carcinogenesis*, it is likely that the level of total caloric intake may be equally or more important. Although many different studies have addressed this factor, three points may demonstrate the range of evidence: 1) rats that have been exposed to a carcinogen develop fewer breast tumors if placed on a restricted-calorie diet; 2) obese human females have increased risk of breast and endometrial cancer; and 3) increased physical activity by humans may reduce the risk of certain cancers.

The causes of cancer are multitudinous: cigarette smoking, certain dietary patterns, radiation, sunlight, occupational hazards, water and air pollution, heredity, and predisposing medical conditions. Given this high level of physiological insult to which the human body is subjected each day, it seems unlikely that one small segment of our diet (polyunsaturated fat) will be shown to play much of a role in carcinogenesis.

Epidemiological data—Information gathered on populations rather than by studying specific individuals or laboratory experiments.

Carcinogenesis—The development of a malignancy or cancer.

Fat and Calorie Reduction in Foods

As discussed in previous chapters, fats and oils play many roles in foods. In 1988, it was estimated that fat contributed 38% of the total calories in the average U.S. diet. In line with the recommendations from governmental and professional groups that this number be reduced to 30% or less, the food industry has sought ways to reduce the fat content of many foods by substituting other materials. This is not an easy task; fat functionality is not readily imitated by other food components.

In addition to decreased dietary fat per se, an overall reduction in caloric intake is desirable for those who are overweight. This entails reducing the caloric density (i.e., Calories per 100 g) of many everyday foodstuffs. Although decreasing the percentage of fat in foods helps, a substantial portion of the components that contribute food energy (i.e., fat, starch, and protein) must be replaced with noncaloric components (*fiber* and water) in order to achieve a significant calorie reduction.

These two approaches tend to be intermingled in the development of tasty, appealing foods that help consumers meet the dietary goals. Of the products recently developed toward these ends, those that have been successful in the marketplace have taken a balanced approach. Fat is reduced by 25–50% (rather than being completely removed), and fiber, gums, and other functional ingredients are used. The products are similar in taste and texture to the ones they replace (in the estimation of consumers) but provide a significant reduction in calories.

Fat Substitutes

FUNCTIONAL EXPECTATIONS

Reduction of caloric density. Simply removing fat from a formula (for example, decreasing fat from 40 to 20% [flour basis] in a cookie dough) reduces the caloric density of the product. However, the weight of fat removed must be replaced with other materials, mainly starch and protein, so the relative calorie decrease is to 4/9 of the original amount, not 0/9. In certain instances, however, nearly all the fat is removed; an example is reduced-calorie white bread, in which the shortening level typically is 0%, although there is a slight increase in emulsifier.

Fat substitutes are often formulated to replace fat on a one-to-one weight basis, where the substitute has a lower calorie count than fat.

In This Chapter:

Fat Substitutes
  Functional Expectations
  Types

Reduced-Fat Products

Reduced-Calorie Products

Fiber—Indigestible or incompletely digested material, usually a complex carbohydrate. Fiber normally contributes bulk but few or no calories to the diet.

For example, several starch-based substitutes are used as gels containing 25% starch and 75% water; 100 g of the gel contributes 100 Cal, and replacing 100 g of fat (at 900 Cal) is a significant calorie reduction. Other substitutes are not used at the one-to-one ratio, but for only partial replacement of fat. Nevertheless, replacing fat (at 9 Cal/g) with a material having fewer Calories per gram results in a lower caloric density in the product.

Mimicking of fat textural contribution. Product texture is the area that presents the most pitfalls in using fat substitutes. Some substitutions work quite well, while others come up short. Some of the *microparticulated protein* products in frozen desserts give a smooth mouthfeel. In baked foods, on the other hand, fat-free (or low-fat) products typically have a dry mouthfeel, even though the actual moisture content may be higher than in the corresponding fat-containing product. It is difficult to predict just how any particular substitute will work in a specific application; the only way to find out is to give it a try.

The problem is that fat is a lubricant during chewing; it aids in clearing food particles from the surfaces (teeth, palate) of the mouth. When fat is not present, the main lubricant is saliva, and if not enough is present, the mouthfeel sensation is one of "dryness." The microparticulated protein replacers provide a "ball bearing" effect in the mouth and hence aid food particle clearance. Unfortunately, during heating (i.e., baking) these proteins are denatured; the ingredient loses its spherical shape and thus its ability to act as a lubricant.

Providing structure. Fat contributes to the structure of many food products. Creme fillings and icings, for example, are made up primarily of sugar and fat, with the fat providing the structure for entrapping air and producing a fluffy, appealing icing. When the fat is removed, other means for stabilizing a foam must be provided. This is usually provided by a gel, formed by a combination of gums, emulsifiers, or proteins with water.

Fat (actually, the oil portion) is a lubricant during the production of some foods, and, in its absence, processing may be rather difficult. This is the case with cookies. Wirecut cookies, for example, are extruded through a die, and dough pieces of the desired weight are cut off by a wire passing across the face of the die. When fat is completely replaced by one of the starch-based fat replacers, the dough sticks on the sides of the die and the resulting dough piece is misshapen. Similarly, rotary molded cookie doughs do not release cleanly from the die molds. In instances such as these, it is necessary to include some amount of oil in the dough, although this amount may be much less than the amount of plastic shortening that is usually used.

TYPES

Table 10-1 lists some materials that have been used as fat replacers. New ones are under development, and this list should not be considered exhaustive. However, it gives an overview of the various types of replacers presently available.

Microparticulated protein— Protein material that has been processed to produce small particles that, when mixed with water, produce a creamy, fat-like sensation in the mouth.

Carbohydrate-based substitutes. Most of the materials listed in the table under this category are used to form a gel containing one part starch and three parts water. The gel is then substituted for fat in the formula on an equal-weight basis; therefore, 100 g of fat (900 Cal) is replaced with 100 g of gel (75 g water, 25 g starch = 100 Cal).

Citrus pectin is actually a gum that forms a gel. Other gums that form gels in water may be used in a similar fashion. Non-gelling gums also contribute viscosity and structure to food products. Gums also are *humectants*, and, because of this water-holding property, they help to overcome some of the dry impression of fat-free foods. A gum is a soluble dietary fiber, so it still must be counted as a carbohydrate (4 Cal/g) in the calculation of calories for labeling purposes.

Humectant—A material that holds water in a finished product.

TABLE 10-1. Fat Replacers

Supplier	Brand Name	Composition
Carbohydrate-based substitutes		
A. E. Staley Mfg. Co.	Sta-Slim	Modified potato starch
A. E. Staley Mfg. Co.	Stellar	Acid-modified corn starch
American Maize-Products Co.	Amalean I	Modified high-amylose corn starch
Avebe America Inc.	Paselli SA2	Potato maltodextrin
Conagra[a]	Oatrim	Oat maltodextrin
FMC Corp.	Novagel NC 200	Microcrystalline cellulose
Grain Processing Corp.	Maltrin	Corn maltodextrin
Hercules, Inc.	Slendid	Citrus pectin
National Starch & Chemical Co.	N-Oil	Tapioca dextrin
Pfizer Chemical Co.	Litesse	Polydextrose
Quaker Oats[a]	Oatrim	Oat maltodextrin
Rhone Poulenc[a]	Oatrim	Oat maltodextrin
Zumbro Inc.[a]	RiceTrin 3	Whole-rice maltodextrin
Protein-based, mixed substitutes		
NutraSweet Co.	Simplesse	Milk, egg white protein gel
NutraSweet Co.	Simplesse 100	Whey protein gel
National Starch & Chemical Co.	N-Flate	Modified starch, gums, emulsifiers, nonfat milk
Low- and noncaloric lipids		
Nabisco	Salatrim	Monostearin esterified with C2:0, C3:0, and C4:0 acids
Procter & Gamble Co.	Caprenin	Monobehenin esterified with C8:0 and C10:0 acids
Procter & Gamble Co.	Olestra	Sucrose hexa-, hepta- and octaesters

[a] Licensee of U.S. Department of Agriculture patented technology.

Polydextrose is made by random polymerization of glucose (dextrose). It is partially metabolized and contributes 1 Cal/g. Polydextrose is usually employed to partially replace sugar in the formulation, but it also has some fat-sparing properties.

Insoluble *dietary fiber*, generally some form of cellulose or hemicellulose, is not usually thought of as a fat replacer. However, it contributes some structure to the product (both during processing and in the finished form), binds water in the finished product (although generally not as much as the soluble gum), and, above all, is excluded from the calculation of calories for labeling purposes. In developing a reduced-fat, reduced-calorie formulation, the inclusion of some insoluble fiber is useful.

Dietary fiber—Fiber in foods that is not digested or absorbed by the body. Insoluble dietary fiber contains no caloric value, while soluble dietary fiber is considered to have some caloric value.

Protein-based, mixed substitutes. A mixture of egg white and milk proteins, subjected to high shear, forms a gel of protein spheroids known as Simplesse. This product is used successfully in frozen desserts. A similar version made from whey protein, Simplesse 100, is approved for use in baked foods.

A blend of several functional ingredients, N-Flate is used as a partial fat replacer in cakes, muffins and cookies. It gives a layer cake, for example, with good volume, grain, and mouthfeel characteristics.

Low-calorie or noncaloric lipids. Extensive esterification of sucrose with fatty acids yields a fat that is not subject to enzymatic hydrolysis in the gut. Therefore, the fatty acids are not freed and absorbed; the ester is simply excreted unchanged. Olestra is the trademark for one such product. Although its safety and physiological reactions were extensively studied, the FDA did not approve it for use in commercial food preparation until early 1996. The idea of synthesizing a lipid material that is not hydrolyzed in the gut has been applied to other starting materials. These have employed a polycarboxylic acid (e.g., citric acid, with three acid groups) esterified to fatty alcohols (e.g. stearol, an 18-carbon alcohol). The specificity of pancreatic *lipase* is such that these ester groups are untouched in the gut. Several other similar lipids have been studied, but none have received FDA approval for food use.

Lipase—An enzyme that hydrolyzes glycerides.

Another approach is to make triglycerides that have the functional properties of ordinary fats but are metabolized somewhat differently, giving a lower caloric yield per gram. Salatrim is monostearin esterified with short-chain acids (acetic, propionic, or butyric). In the gut, the short-chain acids are readily hydrolyzed, absorbed, and metabolized, yielding approximately 4 Cal/g of energy. When stearic acid is released from the monoglyceride, it is only partially absorbed; the calcium stearate soap formed in the gut is largely excreted. The medium-chain fatty acids in Caprenin are hydrolyzed, absorbed, and metabolized, yielding about 7 Cal/g. The monobehenin that remains is not absorbed and is excreted. Studies of animal weight gain with these two triglycerides indicate that each has a biological energy equivalence of approximately 5 Cal/g.

Salatrim (which has several slightly different forms, depending on the ratios of the short-chain fatty acids used to make it) has plastic properties roughly equivalent to those of an all-purpose shortening or a filler fat. It has been used to substitute for these fats in products. Caprenin has an SFI curve like that of a hard butter and has been used as a replacement for cocoa butter in confectionery coatings. Application has been made to the FDA for affirmation of GRAS status for both of these materials.

Emulsifiers. Typically, 2% fat in a formula can be replaced with 1% emulsifier with no loss in functionality. This is not recommended on an economic basis (cost reduction), but it does work when the goal of fat reduction justifies a somewhat higher product cost. Polyglycerol esters contribute about 6 Cal/g. They have been used for years for their emulsifying properties but are also useful for their fat-sparing characteristics. Monoglycerides, polysorbates, and perhaps other emulsifiers can also be used in this way. To illustrate, half of the 100 g (900 Cal) of fat in a cake formula may be replaced with 25 g of polyglycerol ester and 25 g of water. The final caloric contribution is then 600 Cal (9×50 g of fat + 6×25 g of emulsifier), a significant reduction in calories with retention of good product palatability.

Reduced-Fat Products

Salad dressings. While standardized french dressing contains a minimum of 35% vegetable oil (see Chapter 8), a nonstandard pourable dressing may contain any amount of oil, down to 0%. To produce a dressing that clings to salad greens, thickeners such as gums and microcrystalline cellulose are used. The intent is to increase viscosity to give the desired "cling." High-viscosity gums such as xanthan, propylene glycol alginate, and carrageenan are the ones most often used. The gum concentration is typically 0.05–0.3%, depending upon the level of viscosity desired. Since the aqueous phase of pourable dressings is a dilute vinegar, the gums must be resistant to hydrolysis at low pH. Those named above seem best suited to this purpose.

Spoonable salad dressings derive their structure primarily from the modified starch used in their formulation. While the standardized dressing must contain a minimum of 30% vegetable oil by weight, the nonstandard *"lite"* dressings may contain a lesser amount. If the oil is reduced to 15% or less by weight (at least a 50% reduction) the product qualifies for the *"light"* label.

Light (lite)—Form of a product that has 50% less fat than the regular product.

Dairy-type products. Frozen desserts (analogous to ice cream) may have little or no fat. In the simplest traditional sense, a sorbet (a frozen sweetened fruit juice) is such a dessert, but it does not have the creamy mouthfeel associated with ice cream. To achieve this mouthfeel, a combination of gums and a fat replacer is used in place of the fat. The fat replacer that has been most successful in this arena is the microparticulated protein product Simplesse.

Spreads (i.e., margarine and other butter substitutes) have been marketed with a fat content of 40–60% instead of the 80% required for butter. As discussed earlier (Chapter 4, the section on margarines), these are water-in-oil emulsions, stabilized by the formation of a solid fat ma trix during churning or votating. To maintain the integrity of the emulsion at the higher water content, the SFI of the fat phase must be somewhat higher than in butter. In the case of margarine, this is accomplished by selecting basestocks having a higher SFI. However, butter fat is a natural product with an SFI that varies from summer to winter. It can be fractionated to give lower- and higher-melting fractions. By incorporating a higher-melting butter fat fraction into the cream before churning, or by using a blend of butter fat fractions in a process similar to the manufacture of margarine, a reduced-fat butterlike product is produced. For both this "butter" and margarine having reduced fat, the inclusion of flavoring materials in the water phase is important.

Cheese for regular consumption (cheddar, brick, etc.) contains about 33% milkfat. Low-fat versions of these cheeses can be made (by starting with reduced-fat milk containing 1% butterfat), but the texture is generally unacceptable, being tough and "chewy" by comparison with the standard product. A "light" cheese has been developed in which 2% microcrystalline cellulose and 0.6% carrageenan are added to the low-fat milk before the regular cheese-making process is started (1). The final product contains 11% fat, 32% protein, and 49% water, as compared to 34% fat, 24% protein, and 38% water in the control cheese. The carrageenan enters into the *casein*-casein network that is formed during coagulation (initiated by the action of the enzyme chymosin) and softens it. Excess gum interferes with protein coagulation, however, so some of the softening of the final product is supplied by the microcrystalline cellulose. The final product is reported to have eating qualities similar to those of regular cheese and can be processed into other types (e.g., American cheese slices) in the normal fashion.

Casein—The major protein found in milk.

Creme fillings. Many bakery products such as a Swiss roll or a sandwich cookie incorporate a creme filling. This is an aerated mixture of sugar and fat (traditionally butter, but more often an emulsified shortening plus some water and nonfat dry milk). Recently, formulas have

TABLE 10-2. Creme Fillings

Ingredient	Standard	Emulsified	Gum-Stabilized
Sugar	62.8	66.7	68.5
Shortening	23.8	14.0	0.0
Water	3.6	15.0	8.7
Nonfat dry milk	2.0	2.5	4.0
Egg white, liquid	5.0	0.0	0.0
Emulsifier	1.3[a]	1.0[b]	0.0
Gums	0.0	0.0	4.0[c]
High-fructose corn syrup	0.0	0.0	14.5
Salt	0.0	0.3	0.0
Vanilla extract	0.5	0.5	0.3
Calories per 100 g	505	421	332

[a] 0.9% mono- and diglycerides, 0.4% Polysorbate 60.
[b] Decaglycerol dipalmitate.
[c] Blend of gum arabic, alginate, and modified starch.

been developed for making such fillings either with reduced fat or no fat at all. Table 10-2 compares two such fillings with a standard formula for a commercial creme filling. The developer of the "emulsified" filling used decaglycerol dipalmitate, an emulsifier with an HLB of 12, to help stabilize the structure (2). This emulsifier was first dispersed in the water before being combined with the other ingredients. Sugar dissolved in the water phase reduced the *water activity* to the point that the emulsifier gelled, thus contributing some of the structure needed to make a creamy, stable filling. In the "gum-stabilized" filling, *alginate* and the modified starch are gel formers, while the *gum arabic* binds water strongly. Again, the gel provides some of the required structure, but, more importantly, it gives the desired smooth mouthfeel. When a high-moisture filling is used in a sandwich cookie, the water migrates to the basecake, causing it to lose its desired crisp eating quality. The gum arabic binds water, preventing this unwanted event.

Water activity—On a scale of 0 to 1.0, the amount of water in a food that is chemically active and available. Dry foods have a level of 0.1–0.3, and moist foods are in the range of 0.95 or higher.

Alginate—A complex carbohydrate derived from seaweed and used as a thickening agent and gel former.

Gum arabic—A complex carbohydrate derived from Acacia trees and used to increase the stability of emulsions.

Reduced-Calorie Products

Simply reducing the fat content of a food may reduce the caloric density but it often has untoward consequences, as discussed above. Nowadays, consumer demand is not so much for fat reduction per se, but rather for calorie reduction for controlling caloric intake (and help in losing weight). The strategy for developing reduced-calorie foods is somewhat different from that used to develop reduced-fat foods. In the latter case, fat is removed, and then process changes or ingredient additions are made to regain (as much as possible) the textural and mouthfeel contributions made by fat. In calorie reduction, the strategy is to replace caloric components (fat, carbohydrate, and protein) in the finished product with noncaloric (dietary fiber, water) or low-calorie (sugar alcohols, polydextrose) components, while matching the quality characteristics of the product being replaced as nearly as possible. The final formulation and product analysis may be identical; it is just a matter of how development is approached. A few examples will demonstrate the "calorie-reduction strategy."

Reduced-calorie white bread. A typical analysis of white bread is given in Table 10-3. The fat comes from three sources: the flour itself contains about 1.8% fat; shortening is added at 4% of the flour weight; and emulsifiers (monoglycerides, dough strengtheners) contain fatty acids. The first step in reducing calories is to substitute insoluble dietary fiber (α-cellulose or a similar product) for 20% of the flour. The amount of water added to the dough is increased because the fiber absorbs about three times its weight of water. The dough is weak, however, producing a small final loaf, so about 11% *vital wheat gluten* is substituted for an equal weight of flour, necessitating a further increase in dough absorption to 101% (flour basis), as

TABLE 10-3. Reduced-Calorie White Bread

Component	Control		Reduced Calorie	
	Percent	Calories	Percent	Calories
Moisture	38.6	0	45.6	0
Ash	1.8	0	1.6	0
Dietary fiber	1.9	0	11.5	0
Fat	3.8	34	1.4	13
Protein	8.3	33	9.7	39
Carbohydrate	47.6	191	30.2	121
Total calories		258		173

Vital wheat gluten—
Concentrated wheat protein that is extracted from flour and used as an ingredient in baked products to improve the structure and strength of a dough.

compared to 64% for the regular bread. The shortening and monoglyceride are omitted, but the resulting loaf is difficult to slice and is still slightly weak. The addition of 0.5% ethoxylated monoglyceride strengthens the dough and, more importantly, lubricates the slicer blades. Analysis of the final product (Table 10-3) shows a large increase in moisture and dietary fiber and decreases in fat and carbohydrate. The caloric density of the product is decreased by one-third.

Cake doughnuts. During frying, cake doughnuts absorb a relatively large amount of the frying fat, roughly equal to the amount of water evaporated from the batter rings. Thus, the batter as mixed may contain about 13% fat, but the fried doughnut generally contains 22–24% fat. The addition of an ingredient to the batter that holds moisture (retarding evaporation in the fryer) would be expected to decrease fat absorption. This proved to be the case (3). A long-fiber type of α-cellulose, which absorbs about nine times its weight in water, was the most effective. When this fiber was substituted for 3% of the flour in the batter, the fried doughnut contained 18.2% fat (vs. 21.8% in the control doughnut) and 24.4% moisture (vs. 20.9% for the control). Calorie reduction amounted to only 10%, and no further work aimed at this goal was reported.

Snack crackers. Crackers and cookies are low-moisture products, usually containing 2–3% moisture when packaged. Thus, increasing (noncaloric) water in the finished product is not a feasible tactic. Calorie reduction depends on increasing the dietary fiber content and decreasing the fat content. A typical snack cracker contains roughly 11% fat when it exits the oven. It is sprayed with oil, amounting to about 15% of the baked cracker weight; fat analysis of the final product is about 23%, and the cracker has 486 Cal per 100 g. A high-fiber cracker has been developed that contains about 28% total dietary fiber. In developing the formula, it was necessary to add vital wheat gluten to the dough to provide adequate strength for the sheeting process. Because of the gluten and the large amount of fiber, much more water in the dough than normal was necessary, and removing this extra water almost doubled the baking time. The shortening in the dough was halved, so the baked cracker contained about 5.5% fat. Spray oil was applied to get acceptable eating qualities (in particular, a moist mouthfeel and good clearance from mouth surfaces). However, it was found that 5% spray oil was enough to get the desired mouthfeel and texture. The finished cracker contained 10% fat, 28% fiber, and had 320 Cal per 100 g, a one-third reduction from the typical snack cracker.

Oatmeal cookies. A development project for a reduced-calorie oatmeal cookie provides a good object lesson in the overall strategy (4). It is particularly instructive because the necessary sensory evaluation studies are part of the report. The base formula was separated into four segments: flour (30%), sugar (26%), shortening (18%), and other ingredients (oats, whole eggs, leavening, salt, and water). Substitution aimed at lowering calories was made in each of the first three categories. In the sample test formula reported, the following changes were made. The

30% flour component was replaced with 19% flour, 8% α-cellulose, and 3% oat fiber. The sugar component was replaced with 6.5% sugar, 6.5% isomalt (a sugar alcohol derived from sucrose, contributing 2 Cal/g), and 13% polydextrose. The shortening part was replaced with 10% fat, 2.7% mono- and diglycerides, 1% diacetyltartaric ester of monoglyceride (DATEM), 1.6% polydextrose, and 2.7% water. The fourth group was used unchanged. When both formulas were analyzed at 2.5% finished product moisture, the base cookie had 493 Cal per 100 g, while the experimental cookies had 354 Cal per 100 g, a 28% reduction in calories. Analysis of the sensory data showed that the flour and sugar replacements had little or no effect on texture, and the shortening substitutes had the greatest impact on texture attributes.

This study confirms the statement made at the beginning of this chapter: fat functionality is not readily imitated by other food components. Developing a reduced-fat, reduced-calorie product requires using all the ingredients available to the food technologist, in a well-designed experimental plan. The job is not easy, nor is it one of simply replacing fat with a substitute. However, the task is not impossible, and, as time goes on, we can expect to see more high-quality products of this type on the supermarket shelves.

References

1. Bullens, C., Krawczyk, G., and Geithman, L. 1994. Food Technology, 48(1):79.
2. Eydt, A. 1994. Food Technology, 48(1):82.
3. Ang, J. F., and Miller, W. B. 1991. Cereal Foods World 36:558.
4. Campbell, L. A., Ketelsen, S. M., and Antenucci, R. N. 1994. Food Technology, 48(5):98.

APPENDIX A.

Nomenclature and Sources of Fatty Acids

Systematic Name	Common Name	Carbon Length	Source[a]
Butanoic	Butyric	C4:0	Btr
Hexanoic	Caproic	C6:0	Btr
Octanoic	Caprylic	C8:0	Bsu, Cnt, Pk, Btr
Decanoic	Capric	C10:0	Bsu, Cnt, Pk, Btr
9-Decenoic	Caproleic	C10:1	Btr
Dodecanoic	Lauric	C12:0	Bsu, Cnt, Pk, Btr, Ld
5-Dodecenoic	…	C12:1	Sperm whale
9-Dodecenoic	Lauroleic	C12:1	Btr
Tetradecanoic	Myristic	C14:0	Bsu, Cnt, C/S, Plm, S/B, Btr, Ld, Tw
5-Tetradecenoic	Physteric	C14:1	Sperm whale
9- Tetradecenoic	Myristoleic	C14:1	Btr, Ld, mTw
Pentadecanoic	…	C15:0	Ld, mTw
Hexadecanoic	Palmitic	C16:0	Bsu, Cnt, Coc, Crn, C/S, Olv, Plm, Pnt, Saf, Sun, S/B, Btr, Ld, Tw
9-Hexadecenoic	Palmitoleic	C16:1	Cnt, C/S, Olv, Btr, Ld, Tw
Heptadecanoic	Margaric	C17:0	Ld, Tw, mTw
Octadecanoic	Stearic	C18:0	Bsu, Cnt, Coc, Crn, C/S, Olv, Plm, Pnt, Saf, Sun, S/B, Btr, Ld, Tw
9-Octadecenoic	Oleic	C18:1	Same as for C18:0
9,12-Octadecadienoic	Linoleic	C18:2	Same as for C18:0
9,12,15-Octadecatrienoic	Linolenic	C18:3	S/B, Crn, HEAR, LEAR, Btr, Ld, mTw
Eicosanoic	Arachidic	C20:0	Olv, Plm, Pnt, Saf, S/B, Ld, Tw
9-Eicosenoic	Gadoleic	C20:1	Marine oils
5,8,11,14-Eicosatetraenoic	Arachidonic	C20:4	Marine oils
Eicosapentaenoic	…	C20:5	Herring, menhaden, salmon
Docosanoic	Behenic	C22:0	Pnt, HEAR, LEAR
13-Docosenoic	Erucic	C22:1	HEAR
4,8,12,15,19-Docosapentenoic	Clupanodonic	C22:5	Marine oils
4,7,10,13,16,19-Docosahexenoic	Nisinic	C22:6	HEAR, menhaden, salmon

[a] Bsu = babassu, Btr = butter, Cnt = coconut, Coc = cocoa butter, Crn = corn, C/S = cottonseed, HEAR = rapeseed (high erucic acid rape), Ld = lard, LEAR = canola (low erucic acid rape), Olv = olive, Plm = palm, Pk = palm kernel, Pnt = peanut, Saf = safflower, S/B = soybean, Sun = sunflower, Tw = tallow (beef), mTw = mutton tallow.

APPENDIX B.

Composition of Fats and Oils

	Iodine Value	C8:0	C10:0	C12:0	C14:0	C16:0	C17:0	C18:0	C20:0	C22:0	C16:1	C18:1	C20:1	C18:2	C18:3
Lauric fats															
Babassu	13–18	6.0	5.1	42.4	16.8	9.3	...	3.5	0.1	...	...	14.2	...	2.4	...
Coconut	7–12	7.1	6.0	47.1	18.5	9.1	...	2.8	0.1	...	...	6.8	...	1.9	0.1
Palm kernel	14–19	3.3	3.4	48.2	16.2	8.4	...	2.5	0.1	...	...	15.3	0.1	2.3	...
Vegetable oils															
Cocoa butter	33–40	...	...	...	0.1	26.3	0.3	33.8	1.3	0.2	0.4	34.4	0.1	3.1	...
Corn	118–128	...	...	...	0.1	10.9	0.1	2.0	0.4	0.1	0.2	25.4	...	59.6	1.2
Cottonseed	98–118	...	...	0.1	0.7	21.6	0.1	2.6	0.3	0.2	0.6	18.6	...	54.4	0.7
Olive	76–88	...	...	...	...	9.0	...	2.7	0.4	...	0.6	80.3	...	6.3	0.7
Palm	50–55	...	...	0.3	1.1	42.9	0.1	4.6	0.3	0.1	0.2	39.3	...	10.7	0.4
Peanut	84–100	...	...	...	0.1	11.1	0.1	2.4	1.3	2.9	0.2	46.7	1.6	32.0	...
Rapeseed	100–115	...	...	...	0.1	3.8	0.1	1.2	0.7	0.5	0.3	18.5	6.6	14.5	11.0
Canola	100–115	...	...	...	0.1	4.1	0.1	1.8	0.7	0.3	0.3	60.9	1.0	21.0	8.8
Safflower	140–150	...	...	...	0.1	6.8	...	2.3	0.3	0.2	0.1	12.0	0.1	77.7	0.4
(High oleic)	82–92	...	...	...	0.1	3.6	...	5.2	0.4	1.2	0.1	81.5	0.2	7.3	0.1
Soybean	123–139	...	...	...	0.1	10.6	0.1	4.0	0.3	0.3	0.1	23.2	...	53.7	7.6
Sunflower	125–140	...	...	...	0.1	7.0	0.1	4.5	0.4	0.7	0.1	18.7	0.1	67.5	0.8
(High oleic)	81–91	...	...	...	...	3.7	...	5.4	0.4	0.1	0.1	81.3	...	9.0	...
Animal fats															
Butter oil	25–42	1.2	2.5	2.9	10.8	26.9	0.7	12.1	...	...	2.0	28.5	0.1	3.2	0.4
Chicken fat	74–80	...	...	0.1	0.8	25.3	0.1	6.5	0.2	...	7.2	37.7	0.3	20.6	0.8
Lard	48–65	...	0.1	0.1	1.5	26.0	0.4	13.5	0.2	...	3.3	43.9	0.7	9.5	0.4
Tallow (beef)	40–55	...	...	0.1	3.2	24.3	1.5	18.6	0.2	...	3.7	42.6	0.3	2.6	0.7
Fish oil															
Menhaden	159–165	...	...	0.1	10.8	23.2	...	4.2	0.4	0.1	11.4	10.6	1.3	1.8	1.7
Menhaden PHO	78–85	...	...	0.1	10.5	24.1	...	5.2	0.7	0.3	15.0	12.5	4.9	2.4	0.2

[a] Certain fats and oils have significant amounts of fatty acids other than those listed above.
  Butter oil: C4:0, 3.6%; C6:0, 2.2%;C15:0, 2.1%; C14:1, 0.8%.
  Tallow (beef): C17:1, 0.8%.
  Peanut oil: C24:0, 1.5%.
  Rapeseed (high erucic acid rape) oil: C24:0, 1.0%; C22:1, 41.1%; C20:2, 0.7%.
  Canola (low erucic acid rape): C24:0, 0.2%; C22:1, 0.7%; C20:2, 0.0%.
  Menhaden oil: C16:2–4, 4.7%; C18:4, 2.1%; C20:2–4, 3.2%; C20:5, 11.9%; C22:1, 0.2%; C22:4–6, 9.0%.
  Menhaden partly hydrogenated oil (PHO): C16:2, 0.9%; C20:2–4, 10.5%; C22:1, 1.7%; C22:2–4, 7.9%.

APPENDIX C.

Suggested Specifications for Industrial Shortenings and Margarines

Properties Common to All Good-Quality Fats and Oils

Chemical characteristics
  Peroxide value: 1 meq/kg maximum
  Free fatty acid (as oleic acid): 0.05% maximum
  Phosphorus content: 1 ppm maximum

Physical characteristics
  Color: (Lovibond): 1.5 R, 15 Y maximum
  Flavor: Bland
  Odor: Neutral when warmed

RBD (Refined, Bleached, Deodorized) Oil

AOM stability: 10 hr
Oil stability index: 4 hr
Cloud test: 5.5 hr

Light-Duty Frying Oil

AOM stability: 25 hr

Oil stability index: 10 hr

Smoke point: 225°C

Solid fat profiles:

°C	SFC	°F	SFI
10	<5	50	<5
15	<2	70	<1.5
20	<1	80	0
25	0		

High-Stability Frying Oil

AOM stability: 200 hr

Oil stability index: 80 hr

Smoke point: 235°C

Wiley mp: 38 ± 1°C

Solid fat profiles:

°C	SFC	°F	SFI
10	65 ± 4	50	47 ± 3
15	51 ± 4	70	32 ± 3
20	37 ± 3	80	25 ± 2
25	24 ± 3	92	12 ± 1
30	16 ± 2	104	<2
35	7 ± 1		
40	<1		

All-Purpose Shortening

AOM stability: 75 hr

Oil stability index: 30 hr

Wiley mp: 46 ± 1°C

Solid fat profiles:

°C	SFC	°F	SFI
10	38 ± 3	50	28 ± 3
15	29 ± 3	70	20 ± 2
20	22 ± 3	80	17 ± 1
25	17 ± 2	92	13 ± 1
30	12 ± 2	104	7 ± 1
35	8 ± 1		
40	4 ± 1		

Cake and Icing Shortening

AOM stability: 75 hr

Oil stability index: 30 hr

Wiley mp: 46 ± 1°C

α-Monoglycerides:
 3.5% for cake;
 2.5% for icing

Solid fat profiles:

°C	SFC	°F	SFI
10	44 ± 4	50	32 ± 3
15	35 ± 3	70	25 ± 2
20	28 ± 3	80	22 ± 1
25	22 ± 2	92	16 ± 1
30	17 ± 2	104	11 ± 1
35	11 ± 1		
40	7 ± I		

Wafer Filler Fat

AOM stability: 100 hr

Oil stability index: 40 hr

Wiley mp: 35 ± 1°C

Solid fat profiles:

°C	SFC	°F	SFI
10	76 ± 6	50	55 ± 4
15	60 ± 5	70	39 ± 3
20	45 ± 5	80	29 ± 3
25	32 ± 4	92	4 ± 1
30	17 ± 2	104	<1
35	1 ± 1		
40	<1		

Sandwich Cookie Filling Fat

AOM stability: 100 hr

Oil stability index: 40 hr

Wiley mp: 39 ± 1°C

Solid fat profiles:

°C	SFC	°F	SFI
10	53 ± 4	50	38 ± 3
15	39 ± 3	70	24 ± 2
20	28 ± 3	80	18 ± 1
25	21 ± 2	92	9 ± 1
30	11 ± 1	104	<2
35	5 ± 1		
40	<1		

Hard Butter, Coating Fat

AOM stability: 200 hr

Oil stability index: 80 hr

Wiley mp: 38 ± 1°C

Solid fat profiles:

°C	SFC	°F	SFI
10	88 ± 5	50	64 ± 5
15	75 ± 5	70	52 ± 4
20	61 ± 5	80	44 ± 4
25	47 ± 4	92	20 ± 2
30	32 ± 4	100	6 ± 1
35	10 ± 1	104	0
40	0		

Puff Paste Margarine

AOM stability: 200 hr

Oil stability index: 80 hr

Wiley mp: 54 ± 2°C

Moisture content: 18%

Salt content: 2%

Solid fat profiles:

°C	SFC	°F	SFI
10	47 ± 4	50	34 ± 3
15	40 ± 4	70	30 ± 3
20	33 ± 3	80	27 ± 2
25	27 ± 2	92	22 ± 2
30	21 ± 2	104	16 ± 1
35	16 ± 1		
40	11 ± 1		
45	7 ± 1		

General Purpose Margarine

AOM stability: 200 hr

Oil stability index: 80 hr

Wiley mp: 47 ± 2°C

Moisture content: 17%

Salt content: 3.0%

Solid fat profiles:

°C	SFC	°F	SFI
10	38 ± 3	50	28 ± 2
15	30 ± 3	70	21 ± 2
20	23 ± 2	80	18 ± 1
25	17 ± 1	92	15 ± 1
30	13 ± 1	104	10 ± 1
35	10 ± 1		
40	7 ± 1		
45	4 ± 1		

Glossary

AcetylCoA—A metabolic form of carbon in the body that is a basic buildin block for the formation of more complex molecules.

Active filtration—Using pressure to force material through a filter.

Active oxygen method—A procedure to determine how rapidly an oil or fat oxidizes to form peroxides.

Adipose tissue—Tissue in which fat is stored.

Alcoholysis—A chemical reaction in which fatty acids react with alcohol to form an ester.

Alginate—A complex carbohydrate derived from seaweed and used as a thickening agent and gel former.

Alpha (α)-tending emulsifiers—Oil-soluble emulsifiers that form a solid film at the oil-water interface.

Aliphatic—Describing a straight chain of carbons with no branching or ring structure.

Alkalized (dutched) cocoa—Cocoa products that are treated with an alkali agent to raise the pH and produce a darker color.

Amphiphilic—Describing a compound that possesses both lipophilic ("fat-loving") and hydrophilic ("water-loving") regions.

Amphoteric—Describing a compound such as a protein that has both positive and negative charges.

Amylopectin—The type of starch molecule that has branches.

Amylose—The type of starch molecule that occurs as a linear coil without branching.

Animal fats—Fats (like butter, lard, and tallow) derived from animals.

Anionic—Describing a negatively charged compound.

Anisidine value—A test to determine the amount of reaction products produced by lipid oxidation.

Antioxidants—Compounds that can inhibit the development of lipid oxidation.

Atherosclerosis—A disease of the circulatory system caused by clogging of the arteries.

Autocatalytic—Describing reaction that is induced by one if its own products.

Autoxidation—A reaction in which fats undergo oxidative changes due to the double bonds in their structure. The reaction can initiate and proceed without outside influences.

Available heat—The caloric or energy content of food materials, taking into account the amount of material that is absorbed from the digestive tract.

Baker's margarine—A product similar in composition to butter but containing hydrogenated vegetable oil rather than butterfat.

Baker's chocolate—An unsweetened form of chocolate that is the same as chocolate liquor. While having a bitter taste, it is used as an ingredient that adds color and chocolate flavor.

Basecake—The baked product or surface to which a filling or icing is applied.

Basestocks—Fats with certain composition and melting characteristics that are mixed in order to get desirable melting properties in a margarine or shortening.

Bingham plastic—A material that behaves like a solid until it is stirred at a high enough rate so it starts to flow and behave like a liquid.

Bleaching—Removing colored substances from an oil by absorbing them onto a solid material such as clay.

Bloom—A dusty white appearance on the surface of chocolate caused by the formation of certain types of fat crystals.

Bomb calorimeter—An instrument that burns a sample of food and determines how much heat energy it releases.

Calories—The energy contained in food components that gets released and absorbed by the body during metabolism. One calorie is the amount of energy required to raise the temperature of 1 g of water by 1°C.

Capillary melting point—The temperature at which a solid fat turns into a liquid, measured by warming a fat sample in a small tube and observing the temperature at which the solid character of the fat disappears.

Carboxyl group—The chemical functional group on one end of a fatty acid. This is the same as a carboxylic acid (COOH), which can lose a proton and become COO⁻, or combine with an alcohol group to form an ester.

Carcinogenesis—The development of a malignancy or cancer.

Carotenes—A class of fat soluble compounds that are yellow to red in color. Some carotenes, such as β-carotene, are converted to vitamin A in the body.

Carotenoids—Red and yellow coloring agents found in vegetables and grains.

Casein—The major protein found in milk.

Chocolate liquor—The solid mass obtained when the ground, liquified cacao nibs are cooled.

Cholesterol—A fat-soluble compound found in animal products that is required by humans, is produced by the body, and, if present at high levels in the blood stream, is associated with increased risk of diseases of the circulatory system.

Chylomicra—An emulsified form of fat developed in the intestine to transport the absorbed fat in the blood to other organs.

Cloud point—The amount of time an oil remains liquid when cooled to refrigerator temperature.

Cocoa butter—The fat from cocoa beans used in chocolate. It has a sharp melting point just below body temperature.

Cocoa butter substitutes—Hard fat sources, including hydrogenated vegetable oils, that have melting properties similar to those of cocoa butter.

92° Coconut oil—Coconut oil that has been partially hydrogenated to attain a melting point of 92°F (33°C).

Colloid mill—A high-speed grinder that reduces the size of the particles to a fine dispersion.

Compound coatings—Coatings containing fats other than cocoa butter but similar to regular chocolate in melting properties.

Conche—A mixer that slowly mixes a heated paste of chocolate ingredients to reduce the particle size and increase the thickness and smoothness.

Cone stress index—A method of measuring the softness or pliability of a fat.

Conjugated—Describing a situation in which double bonds between carbon atoms occur in a series with one single bond in between (C=C–C=C).

Continuous bread process—A method of dough formation and bread preparation in which ingredients are added, mixed, and fermented without stopping points in the process.

Creaming—1) High-speed mixing of a plastic shortening containing sugar in order to incorporate air. 2) In an emulsion, the collection of the lighter phase in the upper part of the mixture (e.g., oil droplets on top of water).

Cross-linked starch—Starch that is chemically modified by linking the starch molecule chains together laterally with a chemical reagent.

Crumb—The interior of a baked product as distinct from the crust.

Crystals, α, β, β′—When triglyceride molecules in a fat turn from a liquid to a solid as a result of decreasing temperature, they pack into one of three different types of arrangement. Crystal forms exist only when the fat is in the solid state. They can affect the physical properties and functionality of the fat.

Denaturation—Any process in which protein molecules irreversibly change their native shape as a result of forces such as heat, agitation, or solvents.

Deodorization—Removal of odors from an oil by injecting steam. The undesirable odors are carried away because of their volatility.

Dietary fiber—Fiber in foods that is not digested or absorbed by the body. Insoluble dietary fiber contains no caloric value, while soluble dietary fiber is considered to have some caloric value.

Diglyceride—A compound with a glycerol molecule attached to two fatty acids.

Dilatometry—A technique for measuring the amount of solid or liquid in a fat based on small volume changes that occur when going from the solid to the liquid state.

Distilled monoglycerides—A preparation of monoglycerides that is prepared by distillation to separate the monoglycerides from other components, primarily diglycerides.

Dough development—The process of forming gluten from wheat flour with the addition of water and mixing.

Dough strengthener—An ingredient such as a surfactant that is added to a dough in low levels to improve the dough properties.

Dried fish meal—The material remaining when fish tissue is heated to remove the fat and then dried.

Dropping melting point—The temperature at which a solid fat turns into a liquid, measured by warming the fat until it forms a drop of liquid.

Emulsified shortening—A shortening that is manufactured with added emulsifiers (surfactants).

Emulsifier—A material that lowers the interfacial energy between two immiscible phases (e.g., oil and water), thus facilitating the dispersion of one phase into the other.

Emulsion—A homogeneous dispersion of two dissimilar immiscible liquid phases. If oil is dispersed in water, it is an oil-in-water (O/W) emulsion. If water is dispersed in oil, it is a water-in-oil emulsion (W/O).

Energy—The value derived from the digestion and metabolism of food components that is turned into work in a body or tissue.

Enrobing—The process of covering a base food material with a melted coating that hardens to form a solid surrounding layer.

Epidemiological data—Information gathered on populations rather than by studying specific individuals or laboratory experiments.

Essential fatty acids—A class of fatty acids required in the diet of humans. Linoleic and α-linolenic acids are essential fatty acids.

Ester—The chemical linkage that holds an alcohol group (OH) and an acid group (such as COOH) together. An ester bond is the connection between a fatty acid and glycerol in glycerides.

Esterification—The formation of ester bonds by joining hydroxyl groups on glycerol or sugars with carboxylic acids on fatty acids.

Extrusion—The process of putting a product under pressure and sometimes heat and then forcing it out of a barrel through an orifice.

Fatty acids—A group of chemical compounds characterized by a chain made up of carbon and hydrogen atoms and having a carboxylic acid (COOH) group on one end of the molecule. They differ from each other in the number of carbon atoms and the number and location of double bonds in the chain. When they exist unattached to other compounds, they are called free fatty acids.

Fiber—Indigestible or incompletely digested material, usually a complex carbohydrate. Fiber normally contributes bulk but few or no calories to the diet.

Fire point—The temperature at which a heated oil burns with a flame when ignited.

Flash point—The temperature at which a heated oil gives flashes of burning when exposed to a flame.

Flour basis—When determining the percentage of an ingredient in a formula, the weight is compared to the weight of the flour as a percentage. A formula with 50 lb of flour and 5 lb of fat would contain 10% fat on a flour basis.

Foam—A gaseous phase, such as air, dispersed and held in a liquid phase, such as water.

Food additive—A food ingredient category in which certain restrictions on use apply (a regulatory term).

Free fatty acids—Fatty acids with an acid group that is not chemically bound to an alcohol group. Usually fatty acids are bound to glycerol to form triglycerides and are therefore not free.

Free radical—An unpaired electron that is an unstable intermediate in the development of lipid oxidation and rancidity.

Gelatinization—The process in which starch molecules swell and lose crystallinity in the presence of water and heat.

Ghee—The fat remaining when the water and other components are removed from butter. Also termed anhydrous milk fat, it is nearly 100% fat.

Glycerides—Compounds that have one or more fatty acids attached to glycerol.

Glycerol—A three-carbon chain, with each carbon containing an alcohol group. One, two, or three fatty acids may be attached to glycerol to give a mono-, di-, or triglyceride.

Glycerol monostearate (GMS)—A monoglyceride made of one stearic acid molecule attached to glycerol.

Glycerolysis—A chemical reaction in which glycerol is combined with one or more fatty acids to form a glyceride.

Grain (vinegar)—A measure of the strength of vinegar in which 10 grains is equal to 1% acetic acid.

Gum arabic—A complex carbohydrate derived from Acacia trees and used to increase the stability of emulsions.

Gut—General term for the digestive tract or intestinal tract.

Hard butters—Specialty fats that have melting characteristics similar to those of dairy butter or cocoa butter. Used in confectionery products and imitation dairy products.

Hard flake—A hard fat hydrogenated to an iodine value of about 5 and having a melting point of about 57°C (135°F).

HDL—High-density lipoproteins. Molecular complexes found in the blood that carry cholesterol. Cholesterol bound to HDL is being transported and is considered a good type of cholesterol

Helix—A linearly coiled structure with a regular pattern of turns.

Humectant—A material that holds water in a finished product.

Hydrated monoglycerides—A monoglyceride preparation that contains monoglycerides, water, and other surfactants to keep the mixture homogeneous.

Hydrocarbons—Lipid compounds found in trace amounts in fats and oils. They are unsaturated carbon chains such as the compound squalene.

Hydrogenation—The chemical process of adding hydrogen atoms to the double bonds between carbon atoms in a fatty acid. The result is the conversion of a double bond (unsaturated) to a single bond (saturated).

Hydrolysis—A chemical reaction in which a molecule splits into two parts. A molecule of water also splits into H and OH, which are added to the places where the original bond was broken. A fatty acid is removed from a glyceride by hydrolysis of the ester bond.

Hydroperoxide—An intermediate in lipid oxidation in which the fatty acid has added two oxygen atoms and a hydrogen atom at the point of oxidation. It is no longer a free radical but eventually degrades to flavor compounds associated with rancidity.

Hydrophilic—Attracted to water or polar regions of molecules. This chemical property results from the occurrence of oxygen or nitrogen groups and means that there is no attraction to fat or nonpolar groups.

Hydrophilic-lipophilic balance (HLB)—A system of classifying surfactants or emulsifiers by how much "water loving" and how much "fat loving" character the molecules have. On a scale of 0–20, a high number means more hydrophilic.

Hydrophobic—Water-hating or nonpolar.

Hypercholesterolemic—Describing those dietary components that tend to raise the level of serum cholesterol.

Illipé butter—A fat obtained from the seeds of plants grown in tropical Asia; similar to cocoa butter in properties and composition.

Interesterification—Changing the positions of the fatty acids on triglycerides. This is a commercial processing step to change the physical properties of a fat.

Interface—A surface that forms the common boundary between two bodies, spaces, or phases.

Interfacial region—The area where two dissimilar materials are in contact with each other.

Interfacial tension—The forces that cause two dissimilar liquids such as oil and water to separate from each other. **Surface tension** is the force that causes two dissimilar phases such as air and water to separate. While sometimes used interchangeably, interfacial tension is between two liquids or solids while surface tension is between a liquid and a gas.

International Unit (IU)—The amount of a vitamin that exhibits a specific effect when fed to animals.

Intestinal mucosa—Tissue on the walls of the intestine that controls the transport of digested material into the blood stream.

Inverted emulsion—An emulsion in which the continuous phase becomes the dispersed phase and the dispersed phase becomes continuous.

Iodine value—A test to measure the number of double bonds in a fat or oil. A higher value means more double bonds.

Isoelectric point—The pH level at which the number of positive charges is equal to the number of negative charges (e.g., on a protein).

Keyholing—A process in which weak side walls of a bread loaf collapse inward so that the final product is shaped like a keyhole.

Kilocalories—1,000 calories, which is the amount of energy required to raise the temperature of 1 kg of water 1°C. When referring to the caloric content of the diet, the *kilo* part is often dropped and just the term *calories* is used, usually written with a capital C. In the statement that fat has 9 Calories per gram, it means *kilocalories*.

Laminated dough—A dough system that has horizontal layers of dough separated by layers of fat, resulting in a flaky baked product.

Lauric fats—A group of fat sources that are high in lauric acid as a component of the triglycerides.

LDL—Low-density lipoproteins. Molecular complexes found in the blood that attach to cholesterol. Cholesterol bound to LDL is considered bad cholesterol because it deposits on the walls of arteries.

Lecithins—A phospholipid found in egg yolk and soybeans and also used as a food ingredient. It is a surfactant that can stabilize emulsions.

Light (lite)—Form of a product that has 50% less fat than the regular product.

Lipase—An enzyme that hydrolyzes glycerides.

Lipids—A class of compounds found in nature that are soluble in organic (nonpolar) solvents such as ether or hexane. Triglycerides, cholesterol, and vitamin A are examples.

Lipophilic—Attracted to fat or nonpolar regions of molecules. This chemical property results from the occurrence of CH_2 groups and the absence of oxygen or nitrogen groups.

Lipoproteins—Compounds that contain a protein molecule attached to a lipid molecule.

Lubricity—A desirable slippery sensation in the mouth imparted by fats.

Margarine—A product category similar to dairy butter in composition and color. It contains 80% fat, 16% water, and 4% other ingredients such as salt.

Melting point—The temperature at which a solid turns into a liquid. Because they are a mixture of compounds, fats appear to melt over a range of temperature. A specific melting temperature is determined by warming a fat and recording the temperature at which an observable event coinciding with conversion to a liquid occurs.

Metabolism—A series of reactions in the body by which complex food molecules are degraded to simpler forms, energy is extracted, and complex molecules are reformed by the body.

Methylene-interrupted—Describing a situation in which double bonds between carbon atoms occur in a series with two single bonds in between (C=C–C–C=C).

Micelles—Structures in which similar or dissimilar molecules are arranged in an orderly manner.

Microparticulated protein—Protein material that has been processed to produce small particles that, when mixed with water, produce a creamy, fatlike sensation in the mouth.

Milk chocolate—A form of chocolate that has added milk solids, sugar, and other ingredients. It is lighter in color than semisweet or unsweetened chocolate.

Miscella—The mixture of solvent and oil that results from the solvent extraction of oil from oil seeds.

Monoglyceride—A compound with a glycerol molecule attached to one fatty acid.

Monounsaturated—Describing a fatty acid that has one double bond (C=C) in the carbon chain. Oleic acid is the most common of these.

Nibs—The insides or kernels of cacao beans when the hulls are removed.

Nonionic—Describing a compound with no positive or negative charge.

Nonpolar lipids—Fat components that are like organic solvents and not like water in their solubility properties. Unaltered triglycerides are usually very nonpolar.

Nontempering coatings—Coatings that are made of fats that do not readily undergo crystal form changes and therefore do not need tempering to inhibit chocolate bloom.

Obesity—The condition of being overweight due to the accumulation of fatty tissue in the body. Obesity usually is considered being more than 20% over ideal weight.

Oil stability index—An automated procedure for determining the speed at which oxidized products develop in a heated oil when air is bubbled through.

Omega fatty acids—A method of nomenclature that designates the number of carbons between the terminal -CH_3 group and the last double bond in the fatty acid. This is useful is discussing the physiological role of certain polyunsaturated fatty acids.

Organoleptic evaluation—Evaluating quality by using a sense, such as taste or smell.

Oven-spring—The increase in volume that occurs when baked products are first put into the oven. It results from the release or expansion of gases that takes place before the structure of the product is set.

Oxidation—Chemical reaction in which the double bond on a lipid molecule reacts with oxygen to produce a variety of chemical products. The consequences of this reaction are loss of nutritional value and formation of the off-flavors associated with rancidity.

Oxidative rancidity—Off-flavors in a fat or oil due to the reaction of oxygen with fat molecules.

Oxygen bomb—A procedure to measure the speed at which a heated sample consumes oxygen due to lipid oxidation.

Oxystearin—A preparation of oxidized and polymerized hydrogenated vegetable oil that is added to oils to inhibit the crystallization of triglycerides.

Passive filtration—Running material through a filter using gravity.

Peroxide value—A number that indicates the level of peroxides in a fat or oil that has developed as a result of oxidation. Peroxides are considered intermediates in the lipid oxidation reaction scheme.

Peroxides—Oxidized fat molecules that eventually degrades to off-flavors.

Peroxyl radical—An intermediate in lipid oxidation in which the fatty acid radical has added two oxygen atoms and is still a free radical. It is characterized by the structure COO•.

Phenolphthalein end point—A color indicator used to determine when all the acid groups have been reacted with basic groups from sodium hydroxide added during a titration.

Phospholipids—Natural components of fat that have phosphorous associated with the glycerides. Phospholipids are surfactants that assist in emulsification.

Planetary mixer—A mixer with a bowl and paddle arrangement that uses circular motion.

Plaque—A deposit on the inner walls of arteries that grows and eventually causes blockage.

Plastic fat (shortening)—Fat that contains both solid and liquid triglycerides and at room temperature has a consistency that will hold its shape but is soft and pliable.

Plasticity—A physical property of a fat that describes how soft, pliable, and moldable it is at a given temperature.

Polar lipids—Fat components that are more like water and less like fat in their solubility properties. Introduction of oxygen or nitrogen atoms into lipid molecules makes them more polar.

Polyglycerols—A group of emulsifiers that contain polymerized glycerol and various types and amounts of fatty acids.

Polysorbates—A group of emulsifiers that each contain sorbitans, various types and amounts of fatty acids, and polyoxyethylene chains.

Polyunsaturated—Describing a fatty acid that has more than one double bond (C=C) in the carbon chain. Linoleic acid is an example.

Precursors—Compounds that are the basis for the formation of other compounds.

Press cake—The solid residue remaining when cocoa butter is removed from chocolate liquor by pressing.

Press meal—The material remaining when oil is extracted by mechanical means from oil seeds.

Processing—Removing or otherwise extracting a fat or oil from its natural matrix.

Proofing—A step in preparing yeast-leavened products in which the dough is warmed and allowed to rise. It takes place after an initial fermentation and before baking.

Prostaglandins—A group of specialized lipids that play important metabolic roles in humans. They are formed in the body from dietary essential fatty acids.

Protonated—Describing a chemical group that can reversibly add a hydrogen ion (proton) to its structure. If the hydrogen is present, the chemical group is in the protonated form. COOH is the protonated form of COO⁻.

Rancidity—An off flavor in a fat or oil caused either by oxidation or by the release of flavorful fatty acids from the triglyceride.

RBD oil—Refined, bleached, and deodorized oil. These three treatments are frequently applied in series to convert extracted oils into more desirable products.

Reduction—Changing an acid group on a fatty acid to an alcohol group. This is done with metal reducing agents to create fatty alcohols for industrial uses.

Refining—Removing impurities from an extracted fat or oil.

Refractive index—A physical property of a substance that relates to how light is refracted from the material. Usually used to indirectly measure some other property such as concentration.

Rendering, dry— Heating and dehydrating animal tissue to extract and separate the fat. This process results in products with increased color and flavor, so it is not commonly used for edible animal fats.

Rendering, wet—Heating animal tissue by steam under pressure so the fat is released to allow separation.

Retardation of dough—A cooling step that slows down the development of dough and the growth of yeast and allows fats to become more solid.

Reversion flavor—Mild off-flavor developed by a refined oil when exposed to oxygen. Reversion occurs rather easily, and the off-flavor, while undesirable, is not as objectionable as rancidity caused by oxidation.

Rheological—Describing flow properties caused by an outside force or deformation.

Rotary-molded cookies—Cookies formed by pressing dough into a mold engraved on a cylinder.

Salad oil—A refined liquid oil that does not cloud when stored under refrigeration conditions.

Saponification—A chemical reaction caused by addition of alkali in which the fatty acids attached to a glycerol are cleaved off to produce soap (fatty acid salts) and a glycerol molecule.

Saturated—Describing a carbon chain in which the carbons are connected to each other by single bonds, drawn as C–C. It has no carbon-to-carbon double bonds.

Seed crystals—Small pieces of solid fat in a crystalline form that can induce the formation of additional fat crystals upon cooling.

Semisweet chocolate—A form of chocolate with added sugar. The amount of sugar or sweetness can vary.

Sequestrants—Compounds that bind or form complexes with a second compound so that the second compound is no longer chemically active. Positively charged metal ions such as calcium are often sequestered by compounds such as citric acid or EDTA.

Serum—The liquid fraction of the blood that remains after clotting. Cholesterol is measured in the serum portion of the blood.

Shea oil—A fat obtained from tree seeds grown in Africa; similar to cocoa butter in properties and composition.

Shear stress—The application of a force such as stirring on a material.

Sheeting—Stretching dough horizontally to develop the gluten structure.

Shortening—A type of fat used in baking or frying. The name comes from the ability to tenderize or "shorten" baked products.

Slip point—The temperature at which a solid fat becomes more liquid-like. It is measured by warming a fat sample in a small tube and observing the temperature at which it moves.

Smoke point—The temperature at which a heated oil begins to give off smoke.

Solid fat content—A measure of the amount of solid fat in a fat at various temperatures, determined by nuclear magnetic resonance. It is considered a more direct measure than the solid fat index.

Solid fat index—A measure of the amount of solid fat in a fat at various temperatures. It is determined by the volume changes that occur as a result of melting or crystallization.

This index relates the proportion of liquid to solid fractions in a fat.

Specifications—A set of chemical and physical quality requirements that a product or ingredient must meet before it is acceptable.

Spent flakes—The material remaining when the oil is removed from oil seeds.

Stability—The resistance of a fat source to the formation of rancidity.

Staling—The firming of baked products that occurs during storage. The loss of soft texture and flavor is related to changes in the starch after baking.

Sterols—Lipid compounds found in trace amounts that have ring structures rather than the straight chains associated with fatty acids. Examples are cholesterol in animal products and phytosterols in plant products.

Sucrose esters—Emulsifiers manufactured by adding fatty acids to a sucrose molecule.

Surfactant—A chemical compound that concentrates at the interface between two dissimilar phases such as oil and water. The surface tension is lowered by the presence of a surfactant.

Sweet goods—Baked products such as Danish pastries that are high in sugar and often have added icing.

Tallow—A hard white fat obtained from beef or sheep.

Tempering—During the manufacture of shortenings, holding the package product at carefully controlled temperatures to make slight changes in the crystals of fat. This extends the temperature range over which a shortening remains plastic.

2-Thiobarbituric acid value—A measure of lipid oxidation that determines the concentration of certain end products of lipid oxidation.

Thixotropy—A physical property that makes a gel or semisolid turn into a liquid and flow when stirred.

Tocopherol—A class of fat soluble compounds that have vitamin E activity and function as antioxidants.

Totox number—A measure of the total amount of intermediate compounds (peroxides) and end-product compounds that result from lipid oxidation.

Triglyceride—Three fatty acids attached to a glycerol molecule. If the three fatty acids are the same, it is a simple triglyceride; if they are different from each other, it is a mixed triglyceride. Mixed triglycerides are the most common chemical components in fats and oils.

Varnish—An insoluble residue that results from the polymerization and buildup of fat on a hot surface.

Vegetable fats—Fats and oils derived from plant sources.

Viscosity—The thickness of a liquid or semiliquid material.

Vital wheat gluten—Concentrated wheat protein that is extracted from flour and used as an ingredient in baked products to improve the structure and strength of a dough.

Votating—A process that mixes, cools, and whips air or other gas into a fat.

Water activity—On a scale of 0 to 1.0, the amount of water in a food that is chemically active and available. Dry foods have a level of 0.1–0.3, and moist foods are in the range of 0.95 or higher.

Waxes—Lipid compounds that are fatty acids linked to long-chain fatty alcohols. They occur naturally in unrefined oils.

Waxy maize starch—Corn starch that is high in the amylopectin type of starch and low in the amylose type of starch.

Wiley melting point—The temperature at which a solid fat turns into a liquid, measured by warming a fat until it loses its shape.

Winterization—A process in which salad oils are cooled until high-melting-point triglycerides form crystals. These crystals are removed so that the next time the oil is cooled, the cloudiness that comes with crystallization will not occur.

Wire-cut cookies—A type of cookie prepared from a dough in which the individual cookies are cut from the dough piece with a wire before baking.

Yield value—The amount of shear or stirring needed to turn a Bingham plastic from a solid to a liquid.

Index